Mildred Ellen Orton and Vrest Orton

Cooking with Wholegrains

In 1946, Mildred Ellen Orton (1911–2010) and Vrest Orton (1897–1986) opened the Vermont Country Store in Weston, stocking it with products that had largely disappeared from modern stores. Today their son, Lyman, and his sons, Cabot, Gardner, and Eliot, manage the store and a greatly expanded version of their original catalogue, as well as a website, www.Vermont CountryStore.com.

T0039985

LEFT: Vrest and Mildred Ellen Orton circa 1947, standing in front of the stove on which Mrs. Orton cooked for many years.
RIGHT: Mildred Ellen Orton at ninety-eight.

Cooking with Wholegrains

The Basic Wholegrain Cookbook

MILDRED ELLEN ORTON

Foreword by Deborah Madison

Introduction by Vrest Orton

FARRAR, STRAUS AND GIROUX

NEW YORK

Farrar, Straus and Giroux
18 West 18th Street, New York 10011

Printed in the United States of America
Originally published in 1951 by Farrar, Straus and Giroux
First paperback edition, 1971
This paperback edition, 2010

Library of Congress Control Number: 2010926565
ISBN: 978-0-374-53261-1

Designed by Vrest Orton

www.fsgbooks.com
P 1

Contents

Foreword: Good to Be Home

Once in a while I come across an old cookbook, sometimes a much older one, that immediately speaks to me. It may be the subject of the book or the writing that makes me take notice. Sometimes it's the inside of the book that strikes a chord—the illustrations, the typeface, the paper. Or a divine combination of all these elements. Encountering such a book often makes me think, "Oh, I would love to make a book like that!" But such books often lie far from our modern sensibilities. We've become accustomed to slick paper, photographs, lengthy headnotes, and whacky type. Still, there is something to be gleaned from the books of our past, for they often speak with surprising force to the present.

Just such a book is a slim volume entitled *Cooking with Wholegrains: The Basic Wholegrain Cookbook,* by Mildred Ellen Orton, who, with her husband, Vrest, founded the Vermont Country Store. Written in 1951, it was reissued in 1971. The cover of the second book is the color of a loaf of molasses-sweetened (wholegrain) bread. Below Mrs. Orton's name is the promise: "How to Cook Breads, Rolls, Cakes, Scones, Crackers, Muffins & Desserts, Using Only *STONEGROUND WHOLEGRAINS.*" A big topic. A modest book.

Including the introduction by Vrest Orton, the entire volume is only seventy-two pages. His introduction takes up a little less than a quarter of the book, making this really two books in one—the very informative introduction, then the equally appealing, straightforward recipes.

Mr. Orton writes about milling and what lead to the production of refined flours. He's on to the shrewdness of millers and marketers and the nefarious damage they've inflicted on grains since the Industrial Revolution. The demise of nutritious flour is the result of the tail wagging the dog, a story you will enjoy discovering for yourself, so I won't tell it here, but it does explain why grain lost its wholesome quality.

The Italian words for wholegrain bread are *pane integrale*. "Integrity" is not concealed, but contained and visible in this word. Wholeness, integrity, integrated, and integral are part and parcel of the same thing. Wholegrains are complete. Although broken into particles during milling, the parts of the grains are integrated, brought back together, and made whole again. This wholeness is what gives nourishment to our baked goods, and this is something we are still discovering today.

Cooking with Wholegrains first came out about the time my parents were baking sturdy health breads under the guidance of Adelle Davis. Twenty years later the book was reissued, in part because a new need for it was perceived, but also because a fresh generation, capable of taking on wholegrains, had emerged. The flap copy of the 1971 edition says, "Today there is, fortunately, a broader, more intense interest in the subject. A new audience has been born: they are concerned with ecology and a better use of our natural resources, and are discovering that cooking with wholegrains is one good way to restore man's natural relationship with nature."

Coincidentally, I was cooking in a Buddhist community at that time, a community that made up a part of that new audience the Ortons were thinking about. We were taking on wholegrains. We had a stone grinder for grains and a cupboard full of groats and wheat berries, and a baker named Ed Brown came out of that community, who eventually wrote *The Tassajara Bread Book*. Ed's foray into baking wasn't as fiercely focused on wholegrains as this volume is, although they certainly played a big part. Ed was more about doing something wholeheartedly.

Mr. Orton also anticipated a similar approach when he wrote that the best recipes in the world will not make a good bread-maker. Instead, he stressed the place of imagination and intelligence over scripture.

"There is a knack in making good bread," he writes, "which like many other things in life, comes to one by doing after no little trial and error."

He reassures his readers then and now when he says, "It's all in the point of view. Don't make cooking a science. Adopt it as an art"—words so many people today need to hear when they enter their kitchens.

Such thinking, of course, flew right in the face of home economists who, from the late 1900s, had treated matters of the kitchen as a science. Mr. Orton's caution to the reader is a radically different approach. Plus he urges the reader to make good use of one ingredient that is not often called for: one's wit. The generation that embraced wholegrains might call it heart. Same thing.

In contrast to her husband's lengthy introduction, Mrs. Orton's recipes are short, unadorned, and free of seductive headnotes. And because they're expressed with true simplicity, your eye can tell at a glance the nature of a recipe and if it's one that might appeal to you. Maple sugar buckwheat cakes or buttermilk rolls? Wholegrain wheat bread or corn meal dumplings? They all sound tempting. There's even a recipe for the now rare salt-rising bread, which you pretty much have to make yourself if you want to experience the wonderfully odd, cheesy flavor that emerges when it becomes toast.

Most of these charming recipes are for classic American baked goods from New England and, in smaller part, the South. As such they are happily recognizable. Encountering this collection is like running into old friends. You're happy to see them. There's the comfort of your shared pasts and familiar stories. Some you may have forgotten a bit, but you find you can easily renew your acquaintanceships, and when you do, all the goodness of your old friendships resurface.

The recipes in *Cooking with Wholegrains* are like this. They're nourishing. We all know them even if we know only their names. Corn dodgers. (What a great name!) Sour cream biscuits. Boston brown bread. Many of us were raised on baked goods like these, but today they have all been replaced with bakery bread and new kinds of baked goods. Coming to these American classics again, but leaving the white flour behind and replacing it with those big and full grains of our more wholesome pasts, resonates deeply with our need and desire for real nourishment. Foccacia has been fun, but it feels good to be home. Now, once again, is the perfect time for this book.

Deborah Madison, author of *Local Flavors*
and *Vegetarian Cooking for Everyone*

MILDRED ELLEN ORTON

A Word of Warning

In all recipes in this book, to save space and repetition, the grain ingredients are listed simply as wheat flour, oaten flour, soy, rye, corn meal, buckwheat and the like. Naturally we are referring only to 100% stoneground *wholegrains*.

There is some confusion, in the public mind, about two other terms often advertised. One is *Unbleached Flour*. This is simply *white* flour, without the chemical bleach. In no sense is *unbleached flour* a wholegrain because, like any other white flour, it does not contain the natural vitamins and other nutritional elements always present in stoneground wholegrains.

Another term often misused is *Water Ground Meal*. In the beginning stone mills were, by necessity, turned by water power created by a water wheel. *Water* has nothing to do with the meal except to provide power. The important fact is that the stone mill must be turned slowly so as to grind the grains cold, as will be explained in the following pages. A stone mill can be turned slowly by horses, windmills, electric motors, gasoline engines or even atomic energy; it makes no difference what the power is.

And one more word of warning: labels are sometimes misleading on whole wheat products. Federal law requires that all ingredients be listed. Thus under the title *Wholegrain Bread* you can read in very fine print that the ingredients are *wholegrain wheat flour and white flour*. The proportions are never given. They could be one barrel of white flour and one teaspoon of wholegrain flour and be within the law. That's like the classic recipe for Rabbit-Elephant Stew: one rabbit and one elephant!

Mystery of the Mill

> "*And when I have broken the staff of your bread—*
> *Ye shall eat and not be satisfied.*"

I

 COOKING BOOK devoted exclusively to cooking with wholegrain flour is such an old idea that it's brand new. Up to 1850 there was no other kind of cooking book. Yet there is probably not in print today, and certainly there has not been published this side of three-quarters of a century, a basic book of this character. This paradox can be explained by a brief excursion into the history of milling.

But before this excursion into the past, I must define my terms. What is this *wholegrain* that we have all heard so much talk about in the last few years, particularly during the last decade? The answer is simple enough . . . it is *all of* the grain kernel ground into meal or flour, whether it be corn, rye, wheat or oats.

Some confusion arises due to the several names given the *wholegrain* wheat product. In England, it is variously called wholemeal, entire wheat, wholegrain, and whole wheaten flour and in America, Graham flour and "wholewheat" flour. I prefer the term *wholegrain* flour to distinguish from that now prostituted word *wholewheat*, which in America has no more meaning whatsoever due to the shrewd manipulation of the English language by the millers and bakers or their advertising agents. For *wholewheat* is no longer a word meaning *all of* the wheat berry. You can buy "wholewheat" bread and "wholewheat" flour but you are in no sense getting *all of* the kernel of wheat. The millers and bakers get around the federal laws under which manufacturers are supposed to attach true names descriptive of their products by claiming that their product is wholewheat or, in their language, all wheat . . . that is to say, no barley, no corn, no oats. But in no sense is it all *of* the wheat.

Now when and why did the millers, already characterized as shrewd, begin to use less than all of the wheat berry which nature in its munificence furnished to mankind?

II

The story begins in the Stone Age when wheat berries were crushed or pounded between stones by the females of the tribe. After a few thousand years they learned to use saddle-stones, or one concave stone upon which the grain was spread and another stone to rub or grind the grain into a coarse and primitive meal. Even today in Mexico this saddle-stone, called the *metata* there, is used for grinding corn from which the Mexicans make their tortillas. The third and last development of the stone was the *quern* or the stone rotation mill, the first complete milling machine. In existence as early as the second century B.C. this device was composed of two round French Buhr, or granite stones into which grooves were cut (the top turning on the stationary bottom stone), and it persisted down to about 75 years ago in England and America, and indeed is found in back country regions even today.

All of these aforementioned processes obviously ground *all of* the kernel of corn, rye, oats, or wheat into meal, taking no part out and therefore giving real meaning to the word *wholegrain*. Then came the revolution.

About the middle of the 19th Century when the industrial age was reaching out for more production and a mass market, there was invented and developed (chiefly by Hungarians at first and later by other continental Europeans) the roller-mill. The first complete automatic roller-mill was established in the United Kingdom in 1878 at Dublin, although the principle had been introduced into Scotland as early as 1872.

The roller-mill, ancestor of the contemporary milling apparatus in practically all civilized countries, consisted principally of a pair of fluted or grooved metal rollers. Unlike the single pair of cold stones turned slowly by waterpower, these rollers were arranged in a series, beginning with breakers to split the wheat, thus releasing the starch or endosperm so that it might be completely pulverized and ground by the smoother reduction rollers into the white flour we know today. Naturally speed was the main objective. Such a system produced flour just about 100 times faster than the old slow-poke stone mills.

Yet the very slowness of the stone mill constituted its chief value as a food producing tool. The rich germ of the grain kernel has an oil that is susceptible to rancidity when heated in grinding, and if so heated clogs up the grinding surfaces. But the cold stone mills turned slowly, and could pulverize the germ into the flour or meal. But like many another tool of earlier days, character and quality soon became less important than speed, and so the stone mills gave way, about 1870, to another machine.

III

Now we come to the "mystery of the mill." What else did the new high-speed steel roller-mill do except to grind grain 100 times faster? Well, it was soon evident that the germ gummed up the high speed rollers. Therefore, by a series of graduated siftings it was possible to screen out this germ. This epoch-making discovery allowed the millers to expedite their operations, but more significant, they discovered soon enough that flour from which the live and perishable germ was screened out would keep indefinitely on store shelves.

It is rather appalling to realize how easily the milling trade has succeeded these many years keeping these facts from the public, both in America and Great Britain. We only have to look to that famous authority, *The Encyclopedia Britannica*, to obtain bonafide evidence. The erudite and technical account of milling in that compendium was written by an English miller, and it is revealing to see how nonchalantly he treats of the process by which the germ was eliminated. He says, ". . . in roller milling the germ was *easily* separated from the rest of the berry and it was *readily* sifted from the stock. The germ contains a good deal of *fatty matter* which, if allowed to remain, would not increase the keeping qualities of the flour." (The italics are mine.)

That such a standard reference should allow truth to be so distorted is amazing enough, were it not for the more sinister fact that the milling trade has now reached the apotheosis of their craft when they believe their own lies and perpetuate them with no moral guilt whatever.

But it might be interesting to ask what this refined white flour that *keeps* better is doing to our stomachs. No one has summed it up more succinctly than Geoffrey Bowles in that delightful magazine the British *Countryman*, when he declared: "Much of our national illness is caused by crazes for food that is (1) white, (2) refined, and (3) keepable. All three crazes are exemplified in white flour. The best food chemists are the earth and the sun, which produce the wholewheat that the steel rollers of the white flour millers spoil. White flour makes white faces . . . food is stuff to be eaten fresh, not to be 'kept' as if it were an heirloom . . . wholemeal flour naturally does not 'keep' because the germ in it is alive. Germless white flour 'keeps' because it is dead, because it is as dead as Portland cement powder, all its original goodness having been sifted out of it. Let them 'keep' their flour who have no care to keep their health."

IV

Let it not be thought from the foregoing that our cooking book is a polemic against the milling business. But to understand what actually has happened to food during the years of "refinement," it is necessary to look carefully into that simple thing; the wheat berry or the corn kernel, and to understand why it has been treated so unfairly.

Without getting too technical we can say that the wheat berry consists of three sections, (a) the skin or what is commonly known as the bran, which holds the rest of the kernel together, (b) the endosperm, in volume the main part of the berry, consisting mostly of tasteless starch and (c) the germ, embryo, or seed of life from which new life springs if planted, and which gives life-nourishing factors if eaten. This energizing germ contains most of the vitamins and minerals of the berry, very approximately 90%. Yet this germ, in modern milling, is cast aside as a by-product. That such a paradox could persist these many years is astonishing. Yet it is easy to see that when a process like modern milling engages an investment of billions of dollars and results in millions of dollars of profit yearly, the millers are not likely to change in favor of producing a better product when they can make more money with what they are doing.

And it is not because they have been given no opportunity. From a process introduced about 1910, down to the revolutionary milling methods of an engineer named Earle, introduced in England before the Hitler war, millers have been offered machines by which they could not only include the vital germ of the grain in flour, but they could still grind it as fast as by the roller-mill system. Their argument that, to grind the germ into the flour, they would have to return to the stone mill, does not hold water.

The British *Countryman*, at the beginning of the second World War, started a brave crusade for real wholegrain bread in England, but it was not strong enough to beat down the resistance of entrenched millers. In this country there has been no concerted movement to oppose the milling trust, and certainly this author is not foolhardy to undertake what would be at best a quixote adventure. However, this is not to say that in the United States there is not an increasing number of persons discovering the truth about modern milling. And certainly in our small hill village in Vermont we have seen what can happen when intelligent people catch on. The Vermont Country Store stone mill in Weston is now hard at work grinding wholegrain products which are distributed to Americans all over the land by mail.

This wholegrain "revival" is small and will probably never reach more than a million people, because—as authorities have pointed out, there seems to be a perversity in human beings to like things that are not good for them! This is due somewhat to the obdurate fact that the milling and baking groups have for fifty or more years kept alive the public's opinion that refined white flour was refined in more than the technical way, that is, it was socially incorrect to be caught eating dark bread or, as the propagandists now call it, *peasant bread.* Although millions of people today eat what they assume to be wholewheat bread issued by modern bakers, this is in most cases made of nothing more than patent white flour colored with bran.

The original meaning of the word *flour,* from the French *fleur,* signified "the best part," but it apparently is no longer the best part at all. Even the U. S. Department of Agriculture in one of its Bulletins had the courage to state that ". . . experiments have shown that in general the highly refined flours contain practically none of the vitamins present in the whole grain product."

V

A great to-do has been made in recent years over the "enrichment" of bread and flour. In England millers and bakers were compelled by the government to give the people of that country more vitamins per loaf during the war by producing a higher extraction flour, fortified with synthetic vitamins. However, this very slight improvement was fought bitterly by the millers and bakers who played up the horrible fact that the good folk of England were being *forced* to eat peasant bread! Yet in America, these gentlemen eagerly accepted the enrichment idea, as a sop to public demand for better bread.

But this enrichment program is at bottom a delusion and a deceit.*

The original wheat berry contains a number of minerals such as calcium, iron, and phosphorus as well as fat, and the vitamins *A,* *E,* and the *B*-complex group of vitamins. In the wheat berry (as nature grew it) these elements exist in perfect balance and further-

---

*Gayelord Hauser, in his book Look Younger, Live Longer, says, "No vitamin E and only two of the sixteen or more known B vitamins are added to devitalized bread ironically publicized as 'enriched.' What strange mathematics—to take away sixteen and add only two, and call the result 'enriched.' Such misbranding should be dealt with under the Pure Food and Drug Act."

more create a rich nutty flavor which anyone can discover by chewing a few fresh, unground berries of wheat. The "enrichment" of bread consists of putting back into the defunct white flour one or two synthetic vitamins. The *other* vitamins and minerals and nature's balance, as well as the flavor of the original product are omitted. What we really have in these synthetically enriched products is something approaching a patent medicine lost in a welter of starch, and starch, of course is destitute of flavor. The white bread eater walks out of a grocery store with several loaves of white bread under his arm and in order to obtain any discernible nourishment, he must walk across the street to the drugstore and buy several bottles of vitamin pills. But even then he is missing a great deal in actual nourishment and downright eating pleasure. Also, sadly enough, diseases increase and fertility decreases.

Medical discoveries now accepted by all well known nutritionists show that vitamin *E* (which is present in the original wheat berry) benefits many types of heart disease. Reputable physicians after a great deal of research have discovered that the known increase in heart diseases may well be due to the over-refinement of foods such as the removal of the germ, and therefore of the vitamins and minerals from flour.

And when it comes to fertility, it is well established now that the oil in the germ is rich in vitamin *E* which confers upon mankind fertility and vigor.*

This germ is removed by modern milling processes.

VI

It might seem that I have wandered far from cooking. But this volume is an attempt, by the empirical method, to prove that everything good may still be made of *wholegrain* flour in spite of the fact that the hundreds of cook books in print today are based upon germless white flour. The impression has got about (and not un-

---

We have received many letters from people who have benefitted by eating our wholegrain products. The one that pleased me the most was from a lovely girl I know who, after being married for five years without issue, began eating wholegrain at least twice a day, sometimes three. Ten months later I got a letter from her announcing the birth of two bouncing baby boys. Twins!
Of course we can't guarantee this every time.

aided) that you can cook nothing fit for human consumption of wholegrain products and that you need white flour to make beautiful light, fluffy bread, rolls, cakes, etc.

I sometimes wonder just how this food lightness and food whiteness got started. My hunch is that the lighter and whiter a food is, the less tasty and satisfying. I suspect that the desirability of making foods as light as a feather and as white as chalk was brought about, shrewdly, *after* the millers and bakers found by the removal of the germ, and by the introduction of bleaches and chemicals they could make food that, like any other embalmed thing, *kept* until it was sold! They have been able to educate the public into believing that lightness and whiteness are more important than taste and nutrition.

I strongly suspect that modern cooking, promoted by the cook books and cooking schools subsidized by manufacturers of refined and processed products, is a way of making something to eat that will quickly and completely fool the eater into thinking it is something good to eat. One has only to look at the brilliant display of items on the shelves of the supermarkets to realize that the eye-appeal is the main desideratum. I feel that many of us, until we are shown better, are very much like the woman who rushed home from a meeting of the Woman's Sewing Circle and, surrounded by her husband and children, exclaimed with great pride: "Oh, I learned a wonderful recipe for cake today from Mrs. Jones. She gave me her rule for making cake without using *any eggs or butter at all!*" The next step would be cheating at solitaire!

For thousands of years bread was made of genuine *wholegrain* flour, and man thrived. As a matter of fact the highest level of cooking as an art, as well as a method of fortifying the human frame, was reached in the 18th Century. Those great chefs of France, such as Brillat-Savarin and Monsieur Careme now considered the greatest cooks of all time, were plying their art before modern white patent flour was ever dreamt of. They cooked everything with *wholegrain* flour and firmly established cooking among the great arts of civilization. This was a contributing factor in making France one of the great civilized nations of the world.

"COOKING WITH WHOLEGRAINS" is not, however, a book for professional chefs or for those who believe that they can duplicate the success of Brillat-Savarin and Careme. It is for everyday house-wives and also men who are occasionally found in the kitchen, who wish to cook something not only good to eat but good for them!

I will state without fear of contravention that all edibles capable of being cooked with grains can be better cooked with *wholegrains* and, what is more important perhaps, be better eaten for better health and enjoyment.

This book was originally published because of a new concern about nutrition that was being expressed at that time in articles and books by a few pioneer crusaders for better health, such as Adelle Davis. Now that interest has been tremendously intensified not only because many new books now pay attention to good cooking as well as good health, but because the Federal Government, through its several consumer protection agencies, has begun to see to it that the public knows what it is buying from reading the labels on what is being bought.

But today the broader interest in natural foods and wholegrains goes way beyond these factors. A new audience has been born. Young people all over the land are experiencing a new joy that this book reveals. They are discovering that cooking with wholegrains, as a process, is stimulating and rewarding because it produces foods of superlative taste that appeal to an educated and committed gourmet, to whom, of course, good eating is not a necessity but an art.

And the young folk are not inhibited against experimenting. That is why they are finding a new interest in a *Basic Cookbook* like this one which leaves room for the exercise of the imagination and aids in the acquisition of new skills.

Of a necessity, a quite different set of recipes is needed for *wholegrain* cooking than for cooking with germless commercial meals and flour. In using wholegrain products, the relationship between dry and liquid ingredients is different, and other ingredients change in amounts, and especially the knack of combining them. For example, wholegrain corn and wheat contain considerable natural sweetness, and therefore it is never necessary strongly to fortify them with sugar, as must be done when cooking with tasteless and lifeless commercial flour.

It occurred to me that my wife probably knew as much about cooking with wholegrains as any other woman not because of her particular genius as a cook (although she is an excellent one), but because in all these years we have been operating the stone mill here in Vermont she has cooked with wholegrain products!

Being a fellow who believes that things must work to be good, I was forced to the realization that my wife was the only person to prepare this cooking book. So this year she has been hard at work. Every recipe in this book has been cooked, not clipped. Every recipe in this book has been eaten, not sold. Many recipes in this book have been tried time and time again until the right blend was achieved.

I am proud to see that my wife has not committed the cardinal sin of some cooks I have seen and tussled with, and I refer to the sin of using any sugar at all in Southern, or Rhode Island johnny-cake, hoe cake or whatever you prefer to call it. Having eaten and drunk my way pleasantly from New Orleans to Philadelphia, I take off my hat to those lovely ladies of the South, who know by nature that wholegrain corn meal tickles the palate with its own natural sweetness, and that it is as sacrilegious to add sugar to corn dishes as it would be to erect a statue of Sherman in Atlanta.

Further, while those charming ladies of the Providence Plantations still swear by *white* corn meal, which I claim exhibits less than half the flavor and natural sweetness of yellow, I am ready to forgive this fall from grace on the ground that in my early days in their nation I relished with no little gusto the classic Rhode Island johnnycake, flavored with bacon fat, covered with South county butter, and washed down with a beer that was, the last I heard, still made in those parts, and a credit to the country.

My wife won't admit it, but I am of course somewhat of a cook myself! I sneak into the kitchen, particularly in the morning, and knock off muffins or pancakes, at will. However, if I were to write the recipes in this book, they would go something like the one for flapjacks I made last Sunday morning. I give you my recipe for what it may be worth:—

My Own Flapjacks

Dump some wholegrain oaten flour into a large dish and into that some wholegrain wheat flour. Mix these together with the hands so that lumps are broken up. Then shake some baking powder out of the can into the mixture, and some salt. Mix again. In another dish beat up two eggs with wire whip until they run over the edge. Pour this liquid into the dry mixture and then add some milk and stir but not beat. Melt some lard and when cool stir it in. Now you have some flapjacks that when properly and slowly fried on a soapstone griddle on top of a woodburning stove, and then dished onto a plate and covered with butter and maple syrup are fit to eat.

I suppose, however, that it would be difficult for some people to follow such a recipe but frankly I cook by ear. I can never remember how much of anything. If you wish to cook this way, I suggest that you eat a little raw wholegrain meal of different kinds and see how it tastes. And since you know how milk, eggs and other ingredients taste you should be able to put them together in such a way that when they get through everything tastes good. This is *my* method of cooking.

Probably most women will be glad that this is not my wife's.

VII

I want to say a word about bread which my wife has really paid a great deal of attention to in the following recipes. It should be obvious that the bakers' bread we eat nowadays is no longer the staff of life. The most classic evidence of this was reported some years ago in the British Medical Journal. It seems that Sir William Wilcox discovered that in the hospitals of Egypt during the first World War there existed an epidemic of beri-beri which knocked off the white troops but not the Indians. When Sir William looked into the matter he found that the Indian troops had been fed their native flour made of wholewheat ground on stones. The white troops had been treated a little better and fed white flour!

It is too bad that modern penology does not recognize the wisdom of punishments meted out to bread adulterators in the early days. Records show that the making of bread has always been regulated from the Middle Ages and that in the 18th Century it was usual, when bread prices were raised too high, to hang a baker or two. An authority writing of the bakers of Constantinople relates that it was the custom of master bakers to keep a second employee in reserve who in consideration of a small increase in his weekly wage, agreed to appear before the court in case a victim was wanted. In Egypt bakers who sold adulterated bread were nailed to the door posts of their establishments by their ears.

I suppose it is too much to hope that this tradition will be slavishly followed by our government, and as a matter of cold fact, if most of the world wishes to eat sawdust, I don't know as it is any concern of mine. I shall have to be content in the hope that a chosen few who have been introduced to the wholesome, rich, nutty flavor of wholegrains will be now eager to try a hand at new and different ways of cooking with them.

VIII

I would, however, make a final plea for the healthy exercise of the imagination. Cooking, being an art, is no different, at bottom, from any other art. The rules of painting, or the notes of an immortal sonata can be set carefully down, and in fact beautifully printed, but the result is still in the hands, mind, and heart of the practitioner. So with cooking.

The recipes that follow are, like the rules for mixing paints, or the symbols for finding notes on the keyboard, merely to guide you. They are basic recipes, and can, by minor changes, be varied in dozens of different ways. It is your imagination, and your technique (which may be no more than "just getting the right knack") that can help you reach the Olympian heights of all those who practice the culinary arts, where it is realized that we live to eat, rather than eat to live.

VREST ORTON

Bread & Rolls

HE BEST RECIPES in the world will not make a good bread-maker. There is a knack in making good bread, which like many other things in life, comes to one by doing after no little trial and error.

Cooking is not and never will be an exact science, nor indeed a science at all, except as it is practiced by commercial bakers who are able to manufacture, day in and day out, a standard product . . . nearly worthless as food, of course, but always durable and very pretty to look at.

Making good bread is an art! Besides dealing with live factors like the germ-content wholegrain flours, butter, eggs, fresh milk and so on, you are also dealing with temperature and humidity, and last but not least yourself. Bringing all these together, precisely in the same way, upon each occasion, time and time again, cannot be done and in fact ought not to be done.

If it were possible, and you could do so, cooking would lose the abiding pleasure that now lurks in it, and you would become sick and tired of the whole thing.

This is not a sermon, but only a warning. After you have followed a recipe for bread religiously, and it doesn't "come out" right, don't despair or suspect that the recipes are wrong. Try again.

And then, some bright Tuesday morning, you will enter the kitchen, take out this book, put some things together, and bake a bread that is so wonderful you wonder how it ever happened.

Once you do this, you will have the knack. Like learning to ride a bicycle, you can keep it with you as long as you like.

It's all in the point of view. Don't make cooking a science. Adopt it as an art. And be sure you consider another ingredient that is not mentioned anywhere in this book . . . and make use of it often. I refer to your wits.

V. O.

Wholegrain Wheat Bread

3 CUPS WARM WATER	8 CUPS WHEAT FLOUR
2 PACKAGES DRY YEAST	½ CUP BROWN SUGAR
I TABLESPOON BROWN SUGAR	2½ TEASPOONS SALT
I CUP POWDERED MILK	½ CUP LIQUID SHORTENING

In this and all other recipes in this book, all measurements given are level measurements.

This recipe makes four loaves of bread.

YEAST MIXTURE. Dissolve 1 tablespoon of brown sugar in the 3 cups of warm water. Add the packages of yeast and let this mixture stand while you are preparing the flour mixture.

FLOUR MIXTURE. Sift the flour and, if you so desire return the bran siftings to the sifted flour. Now, in a separate bowl, mix together 6 cups of the sifted flour with the powdered milk, ½ cup of brown sugar and salt.

Stir liquid to dissolve the yeast and add half of the flour mixture to this warm liquid and mix thoroughly. Add the liquid shortening (butter, margarine or salad oil may be used) and mix well again.

Add the remaining dry mixture, one cup at a time and mix thoroughly after each addition. Add two more cups of the sifted flour in small amounts as needed to make a stiff dough. When it becomes difficult to handle the mixture with a spoon, empty the dough onto a pastry board and kneed in more flour until the mix-ture is firm, yet light. It may be necessary to use an additional cup or more of sifted flour.

Place the mixture in a large clean, oiled bowl, turning the mixture over in the bowl to coat the dough with oil. Cover the bowl and let rise at about 90° in a slightly warm oven for one hour.

Punch down the dough and turn onto a floured pastry board. Knead well and divide into four equal parts. Knead each section and shape into a loaf. Place each loaf in a buttered pan 8½″ x 4½″ x 2¾″ deep. The pans should be about half full. Each loaf may also be shaped into a round loaf. Then place two loaves in each end of a large pan 9½″ x 5¼″. These loaves may be separated after they are baked.

Cover bread with a cloth and place in a warm oven (about 90°) for 30 minutes. Remove from oven. Bring the temperature to 400° and bake the bread for 15 minutes. Reduce temperature to 350° and bake ½ hour longer. Remove bread from pans immediately. Place on wire racks and butter top of loaves.

Since you may save some of the loaves for later eating it is wise to freeze immediately after baking any surplus loaves because they can dry out very quickly. They may also be stored in the refrigerator for a few days without losing their freshness. To restore the bread to its newly baked freshness and taste, re-heat the bread in a 350° oven for about 15 minutes.

Sour Dough Bread

SOUR DOUGH PRIMER has to be made first, as follows:

2 CUPS WARM WATER

1 PACKAGE YEAST

2 CUPS WHEAT FLOUR

In a medium size bowl, dissolve yeast in water. Mix thoroughly and add the wheat flour, including the siftings. Beat one minute. Let stand, covered, at room temperature for 48 hours to allow fermentation to take place. You now have a sour dough primer from which you use one cup to develop a batch of sour dough bread, returning one cup of bread mixture to the primer, as explained in the following recipe. Store primer in a covered jar in the refrigerator after it has reached the sour stage.

THE BREAD RECIPE is as follows:

3 CUPS WARM WATER	I CUP POWDERED MILK
I PACKAGE YEAST	2 TABLESPOONS BROWN SUGAR
7 TO 8 CUPS WHEAT FLOUR	2 TEASPOONS SALT
I CUP SOUR DOUGH PRIMER	

In a large bowl, dissolve yeast in water, mix, add 3 cups sifted wheat flour, including siftings, and 1 cup of sour dough primer. Beat one minute. Cover and let stand at room temperature overnight, or until desired sourness is reached. In the morning (or when ready to continue) remove one cup of this bread mixture and return to the jar of sour dough primer in the refrigerator.

Now, mix together the powdered milk, brown sugar, salt and 2 cups sifted wheat flour including the siftings. Add to bread mixture and beat well. Keep adding sifted wheat flour gradually. Mix with a spoon as long as possible, then knead in wheat flour until a firm, smooth, elastic dough is formed. Put into a clean, buttered large bowl, turn dough over to coat with butter. Cover and let rise in a warm place about 1 hour or until double in bulk. Dough should be light and soft.

Punch down, knead and shape into four loaves. Place in greased pans. Brush over with melted butter or salad oil. Cover with a towel and let rise about ¾ hour in a warm place until double in bulk and light and soft.

Bake at 400° for 45 minutes. Cool on wire racks.

Salt-rising Bread

I CUP MILK	7 TABLESPOONS CORN MEAL
I TABLESPOON BROWN SUGAR	I TEASPOON SALT

Scald the milk and add the sugar, corn meal and salt. Put in a covered jar and place in a dish of water as hot as the hand can bear. Keep in a warm place overnight. By morning the mixture should show fermentation and gas can be heard to escape. Then add

2 CUPS SIFTED WHEAT FLOUR	2 TABLESPOONS BROWN SUGAR
2 CUPS LUKEWARM WATER	3 TABLESPOONS MELTED SHORTEN-ING

Beat this mixture thoroughly, place in a dish of warm water again and let rise until light and full of bubbles. Then add about 4½ cups of sifted wheat flour, or enough to make a stiff dough. Knead for ten or fifteen minutes, then mold into loaves. Place in greased pans and let rise again until light.

Bake 15 minutes at 425°, then lower the temperature to 375° and bake about 30 minutes longer.

Colonial Cottage Loaf

1 CUP MILK SCALDED	1 PKG. YEAST
1 CUP HOT WATER	¼ CUP LUKEWARM WATER
3 TABLESPOONS SHORTENING	1¾ CUPS CORN MEAL
2 TABLESPOONS MOLASSES	1¾ CUPS WHEAT FLOUR
2 TEASPOONS SALT	1¾ CUPS RYE

Pour the scalded milk and hot water over the shortening, molasses and salt and stir until the shortening is dissolved. When lukewarm add the pkg. of yeast dissolved in the quarter cup of lukewarm water. Sift the corn meal, wheat flour and rye separately, then measure and sift all together into the liquid. Stir with a spoon. The dough should be just soft enough to beat with a spoon.

Beat this mixture for 3 to 5 minutes or until you get tired of doing it longer. Then set in a warm place to rise for about 2 hours. When light and about double in bulk, beat it again. Do this four different times to create a light, soft, spongy dough. Turn onto a board lightly sprinkled with wheat flour or some of the mixed flours. Knead until light (about 5 minutes) then shape into a round loaf and place in a 1 quart greased casserole. Let rise again until about double.

Bake for about 45 minutes, starting at 425° and lowering to 375° after 15 minutes. This makes a delicious, nutty flavored bread.

Rye Bread

2 CUPS MILK SCALDED   1 PKG. YEAST DISSOLVED IN ¼ CUP
1 TABLESPOON SHORTENING   LUKEWARM WATER
2 TEASPOONS SALT   4½ CUPS RYE

Scald the milk with the shortening. Add salt and when lukewarm add the dissolved yeast.

Sift the rye, measure and sift into the liquid. You should add enough rye to make a stiff but sticky dough. It should be moist enough to stir with a spoon. Continue stirring for 4 or 5 minutes the way you do when folding beaten egg whites into any mixture.

Cover closely and set in a warm place to rise until light (about 2 hours). Punch down and turn onto a board lightly sprinkled with rye. Knead for about 10 minutes, or until the dough begins to feel springy. Mold into loaves, cover, set in a warm place and let rise to about double. Bake 300° for 1½ hours with a pan of hot water placed in the oven underneath the bread.

Pumpernickle

1 PKG. YEAST   1 CUP SOUR DOUGH PRIMER
2 CUPS WARM WATER   (FROM PAGE 19)
2 CUPS WHEAT FLOUR   3 CUPS RYE FLOUR
2 TEASPOONS SALT   2/3 CUP POWDERED MILK

Dissolve yeast in warm water, add sifted wheat flour, with siftings included, and beat one minute. Add sour dough primer. Beat ½ min. Take out one cup of this mixture and return to your sour dough primer storage supply. If you don't wish to keep primer on hand, reduce above amounts to 1 cup water and 1 cup wheat flour.

Mix together salt, powdered milk and 2 cups sifted rye flour including siftings. Add to dough mixture, mix well and beat ½ min. Then add one cup or more of rye flour gradually as needed. When too thick to mix with a spoon, knead in more rye flour until a firm dough is formed. Use rye flour sparingly for kneading. Do not overknead. Place in clean, buttered bowl, let rise in a warm place about one hour. Rye flour is heavy and does not rise quite to double in bulk.

Punch down, knead quickly and shape into two loaves. Place in buttered 1 qt. casseroles, or 4 x 7 ½ in. bread pans. Brush tops with salad oil. Cover with a towel and let rise in a warm place about 1 hour.

Bake at 325° for 55 min. with a pan of hot water in bottom of oven. Do not pre-heat oven. Allow bread to rise as oven heats.

VARIATION: 1 teaspoon Caraway Seed may be added to the dough before shaping into loaves, if desired.

Rye and Wheat Bread

3 CUPS LUKEWARM WATER	3 TEASPOONS SALT
2 PKGS. YEAST	5 CUPS RYE
3 TABLESPOONS BROWN SUGAR*	3 CUPS WHEAT FLOUR
3 TABLESPOONS MELTED SHORTENING	

Dissolve yeast in water. Add brown sugar, shortening and salt.

Sift the wheat flour and measure. Stir the rye and measure, then sift the wheat and rye all together and add about ¾ of this mixture to the liquid. Mix thoroughly, then add another cup of the flour, or enough more to make the dough stiff enough to knead.

Cover with a damp cloth, set in a warm place and let it rise until light. Punch down and let rise four different times until dough becomes light and spongy. Then empty onto board and knead for a few minutes. Divide into two equal parts and knead and mold into loaves. Place in greased pans, cover with a damp cloth and let rise until light (nearly double in bulk) and bake 1½ hours at 300° with a pan of hot water in the oven underneath the bread.

Oatmeal Bread

2 CUPS WHEAT FLOUR	2½ TEASPOONS SALT
2 CUPS SCOTCH OATMEAL	4 TABLESPOONS BROWN SUGAR
1½ CUPS MILK	2 TABLESPOONS SHORTENING
1 PKG. YEAST	(MELTED)
¼ CUP LUKEWARM WATER	

To make one loaf, pour scalded milk over Scotch oatmeal and mix well. Add pkg. of yeast dissolved in lukewarm water, then salt, brown sugar and melted shortening.

*Molasses, maple syrup or honey can be substituted for brown sugar in this and other recipes.

Sift the wheat flour, measure, and add to the batter. Mix until all the flour is dampened, then beat with a spoon for 2 minutes. The dough should be thick enough so it is just possible to beat with a spoon.

Cover with a damp cloth, set in a warm place and let rise until light and soft when touched with the finger.

Punch down, cover with a damp cloth and let rise again. Then punch down again before turning out on a board sifted over with wheat flour. Knead 5 or 10 minutes. By this time the dough should not stick to the board at all.

Form into a loaf and place in a greased pan. Cover with a damp cloth, let rise to about double and bake 1 to 1½ hours at 300° with a pan of hot water placed in the oven underneath the bread.

French Brioche

1 PKG. YEAST	¾ POUND BUTTER
¼ CUP LUKEWARM WATER	2 TEASPOONS SALT
4 CUPS WHEAT FLOUR	2 TABLESPOONS SUGAE
7 EGGS	

Dissolve the pkg. of yeast in lukewarm water, then add enough wheat flour to make a very soft ball of paste. Drop this ball into a bowl of warm water (the water must not be hot or it will kill the yeast plant). Cover, and set in a warm place to rise. The ball of paste will sink to the bottom of the water at first, but will rise to the top later, and be full of bubbles.

Put the rest of the flour in a large bowl, make a hole in the center of it and into this hole put the softened butter, salt, sugar and 4 eggs. Break the eggs in whole. Work together with a spoon gradually working in the flour. Add 2 more eggs, one at a time, as you need more moisture. Beat this with a spoon for 5 minutes or so.

When the ball of paste has risen to become light and full of bubbles, lift it out of the water with a skimmer and mix into the dough, adding the last egg. Beat this mixture for a long time. The longer it is beaten, the better and finer will be the grain.

Put the dough in a bowl, cover and let rise to double its size. Beat it down again and place in the refrigerator for at least 12 hours. After removing from the refrigerator, handle the dough delicately and quickly as it softens as it becomes warm. This dough may be shaped into braided rolls, or buns of any desired shape, or tiny loaves of bread. If a glazed top is desired, brush over with egg yolk diluted with water before putting in the oven. Always serve hot or very fresh.

Bake at 425° about 20 minutes.

Buttermilk Rolls

1 CUP BUTTERMILK, WARMED	1 PKG. YEAST
3 TABLESPOONS SHORTENING	2¼ CUPS WHEAT FLOUR
1 TEASPOON BROWN SUGAR	1 TEASPOON BAKING POWDER
¼ TEASPOON SODA	1¼ TEASPOONS SALT

Dissolve the yeast in the warm buttermilk then add the melted shortening, brown sugar and soda and mix well. Sift the wheat flour, measure, add baking powder and salt and sift into the liquid all at once. Mix until the flour is all dampened then beat with a spoon for one minute. Cover with a damp cloth, set in a warm place and let rise to about double in bulk.

Punch down, let rise once more and turn onto a board covered with sifted wheat flour and knead until light and spongy (between 5 and 10 minutes). Mold into any desired size or shape, cover with a damp cloth again and let rise until light. Bake 15 to 20 minutes at 425°.

Wheat Rolls

2 CUPS SCALDED MILK	1 PKG. YEAST
4 TABLESPOONS BUTTER OR SHORT-	¼ CUP LUKEWARM WATER
ENING	2 EGGS WELL BEATEN
2 TABLESPOONS BROWN SUGAR OR	5 CUPS WHEAT FLOUR (APPROXI-
HONEY	MATELY)
2½ TEASPOONS SALT	

Scald milk with shortening, then pour over sweetening and salt. When lukewarm add the yeast dissolved in the lukewarm water and after that the well beaten eggs.

Sift the wheat flour, measure and sift into the liquid using about 4 cups to start with. Mix well and add enough more wheat flour to make a dough as stiff as you can have it but still be able to beat with a spoon for two minutes.

Cover closely and allow to rise in a warm place until almost double. Punch down and let rise again. The punching down process is important as it prevents stretching of the dough.

Knead until light and elastic. Mold into rolls of any desired shape. Cover with a damp cloth and let rise until light. Bake at 425° 15 to 20 minutes. This recipe makes 2 dozen rolls.

Vermont Wheat Rolls

I CUP MILK, SCALDED	I PACKAGE YEAST
2 TABLESPOONS BUTTER	¼ CUP WARM WATER
2 TABLESPOONS SUGAR	I TEASPOON SUGAR
¾ TEASPOON SALT	2½ CUPS WHEAT FLOUR

Scald milk and pour over butter in a large bowl. Add sugar and salt. Stir to dissolve. In the meantime dissolve 1 teaspoon sugar in ¼ cup of very warm water, then add the package of yeast. Let stand until yeast has foamed up nicely. Stir and add yeast to the milk mixture. Mix well. Sift the flour, measure 2 cups and add to liquid. Mix and beat for a minute or two, then add the last ½ cup of flour, a little at a time until a soft ball of dough forms in the bowl. It may require a little more or less than the ½ cup. Butter sides of bowl. Cover and let rise in a warm place for 45 minutes. Punch down. Divide dough into four equal parts, then divide each fourth into three parts. Roll each small portion in floured hands for a few seconds until firm and light. Put into a buttered muffin pan. Cover with a towel and let rise in a warm place 30 minutes. Bake at 400° for 20 minutes. Makes twelve rolls.

Special Recipes
Using Muffin Meal

Muffin Meal is an amalgamation of wholegrain wheat, corn and rye mixed together and provides a basic and simple grain product that can be used in hundreds of ways. A few typical examples are given below and your own ingenuity can provide many others.

Muffins

2 EGGS
I CUP MILK
6 TABLESPOONS MELTED SHORTEN-
  ING
I ½ CUPS MUFFIN MEAL

2 ½ TEASPOONS BAKING POWDER
2 TABLESPOONS BROWN SUGAR OR
  HONEY
¾ TEASPOON SALT

Beat the eggs. Add milk, sweetening and melted shortening. Sift the Muffin Meal, measure, add baking powder and salt. Sift this into the liquid and stir only enough to dampen the meal . . . do not beat. Let stand a minute or two while greasing the muffin pans. Drop mixture by spoonfuls into the muffin pans and bake at 425° for about 20 minutes.

Molasses Quick Bread

I EGG BEATEN	½ TEASPOON SODA
½ CUP MOLASSES	I TEASPOON BAKING POWDER
I CUP BUTTERMILK	I TEASPOON SALT
¼ CUP MELTED SHORTENING	½ CUP SEEDLESS RAISINS
2 CUPS MUFFIN MEAL*	

Beat the egg, add molasses, buttermilk and shortening and mix. Sift the Muffin Meal, measure, add soda, baking powder, and salt and sift into the liquid. Pour raisins on top of flour mixture and stir up quickly all at once until well mixed. Bake in a bread pan at 350° for 45 minutes.

Steamed Brown Bread

3 CUPS MUFFIN MEAL*	2 CUPS SOUR MILK
2 TEASPOON SODA	2 TABLESPOONS MELTED SHORTENING
I TEASPOON SALT	½ CUP RAISINS
2/3 CUP MOLASSES	

Mix together the sour milk and molasses. Sift the Muffin Meal, measure, add soda and salt, then sift into the liquid. Pour raisins on top of flour mixture and mix all together. Add shortening last.

Pour into a greased mold, having the mold no more than ⅔ full, cover closely. Place in a dish of hot water, having the water come half way up on the mold. Steam 3½ hours. If placed in several small molds, steam 2 hours.

Griddlecakes

2 EGGS, BEATEN	2 CUPS MUFFIN MEAL*
I ½ CUPS MILK	3 TABLESPOONS SUGAR OR HONEY
¼ CUP MELTED SHORTENING	¾ TEASPOON SALT
	5 TEASPOONS BAKING POWDER

Beat eggs, add milk, sweetening and melted shortening, and mix well. Sift Muffin Meal, measure, add salt and baking powder. Sift into the liquid ingredients. Mix up very quickly, stirring only enough to dampen the meal. Cook slowly on moderately hot griddle until golden brown and serve with Vermont maple syrup, honey, or jam.

See page 27.

Jam Hot Cakes

2 or 3 tablespoons of jam or jelly used in the above recipe in place of the sugar or honey makes an interesting new kind of hot cake which can be eaten best with butter between the cakes piled at least three high, and maple syrup poured over all.

Buttermilk Griddlecakes

I EGG, BEATEN	I CUP MUFFIN MEAL*
I CUP BUTTERMILK	½ TEASPOON SODA
2 TABLESPOONS MAPLE SYRUP OR HONEY	I TEASPOON BAKING POWDER
4 TABLESPOONS MELTED SHORTENING	¾ TEASPOON SALT

Beat the egg, add buttermilk, sweetening and shortening. Sift the Muffin Meal, measure, add soda, baking powder and salt and sift into the liquid. Mix up quickly, stirring only enough to dampen the meal. Cook on greased hot griddle until golden brown and serve with Vermont maple syrup or honey.

Hot Bread

1 ⅛ CUPS MUFFIN MEAL	I CUP MILK
I TEASPOON CREAM OF TARTAR	4 TABLESPOONS SUGAR OR HONEY
½ TEASPOON SODA	4 TABLESPOONS SHORTENING,
¾ TEASPOON SALT	MELTED
	I EGG BEATEN

Mix together the beaten egg, milk, sweetening and shortening. Sift the Muffin Meal, measure and add other dry ingredients. Sift into the liquid. Beat up quickly, pour into a greased square pan and bake at 425° for about 20 minutes.

Buttermilk Quick Bread

1¾ CUPS MUFFIN MEAL*	¾ TEASPOON SALT
I TEASPOON CREAM OF TARTAR	I CUP BUTTERMILK
½ TEASPOON SODA	2 EGGS WELL BEATEN
2 TABLESPOONS BROWN SUGAR OR HONEY	3 TABLESPOONS MELTED SHORTENING

Beat the eggs, and add buttermilk, melted shortening and sugar or honey. Sift the Muffin Meal, measure, add other dry ingredients, then sift into the liquid. Beat up quickly until free of lumps. Pour into a greased square pan and bake at 425° for about 20 minutes.

See page 27.

Quick Breads, Steamed Breads & Doughnuts

Countryman's Corn Bread

(With sour milk or buttermilk)

1 CUP CORN MEAL	¾ TEASPOON SALT
1 CUP WHEAT FLOUR	1 CUP SOUR MILK OR BUTTERMILK
1 TEASPOON CREAM OF TARTAR	2 EGGS WELL BEATEN
½ TEASPOON SODA	3 TABLESPOONS MELTED SHORTEN-ING
2 TABLESPOONS BROWN SUGAR	

Beat the eggs. Add sour milk or buttermilk. Sift the wheat flour, measure, add all other dry ingredients and sift into the liquid. Add melted shortening, mix all together and beat for a second or two.

Pour into a greased 8 x 8 inch pan, or a pan 1½ inches deep and bake about 20 minutes at 425°.

Corn Meal and Wheat Quick Bread

⅔ CUP CORN MEAL

⅔ CUP WHEAT FLOUR

I TEASPOON CREAM OF TARTAR

½ TEASPOON SODA

I CUP MILK

¾ TEASPOON SALT

4 TABLESPOONS BROWN SUGAR

4 TABLESPOONS SHORTENING

I EGG BEATEN

Cut the shortening into the combined dry ingredients with a pastry cutter. Add the egg and milk beaten together. Give a few quick beats with the spoon, pour into greased 8 x 8 inch pan and bake at 425° about 20 minutes.

Early American Hot Bread

⅔ CUP CORN MEAL

I EGG

I CUP MILK

2 TABLESPOONS MAPLE SYRUP OR HONEY

¾ CUP WHEAT FLOUR

2½ TEASPOONS BAKING POWDER

I TEASPOON SALT

3 TABLESPOONS MELTED SHORTENING

Beat egg until light. Add milk and Vermont maple syrup. Mix together the dry ingredients, and sift into the liquid. Add melted shortening. Stir briskly and bake in 8 x 8 inch pan for 20 minutes at 425°. This hot bread is grand provender for growing kids to thrive on.

Southern Corn Bread

2 CUPS MILK

I CUP CORN MEAL

I TEASPOON SALT

2 TABLESPOONS BUTTER

2 EGGS, WELL BEATEN

I TEASPOON BAKING POWDER

Scald corn meal and milk together. Add the butter. Beat with an egg beater until smooth, cool slightly, and add eggs, baking powder and salt. Bake in greased 8 x 8 inch pan 20 to 25 minutes at 425°.

Sugar must never be used in Southern corn bread. The natural sweetness of the corn is sufficent.

Spoon Bread

I CUP CORN MEAL	2 EGGS WELL BEATEN
1 ½ CUPS BOILING WATER	½ TEASPOON SALT
1 ½ CUPS MILK	½ TEASPOON BAKING POWDER

Stir corn meal into boiling water. Remove from fire and add the milk slowly, then eggs, salt and baking powder. Bake the mixture in a deep buttered pan or casserole for a half hour or longer at 425°. Don't be afraid to set this dish right on the table in the pan or casserole in which it is cooked. Serve smoking hot by the tablespoonful from the pan with a pat of butter tucked into its middle.

Suet Corn Bread

(*sour milk*)

I EGG BEATEN	½ TEASPOON SODA
I CUP SOUR MILK OR BUTTERMILK	¾ TEASPOON SALT
1 ¼ CUPS CORN MEAL	⅓ CUP CHOPPED SUET

Beat the egg and add the sour milk or buttermilk and mix well. Crush soda in the palm of hand to remove lumps and add it, with salt, to the corn meal and mix with a spoon. Add the chopped suet to this corn meal mixture and mix until crumbly, then combine the wet and dry ingredients. Mix well and bake at 425° for 20 minutes.

Quick Brown Bread

I EGG BEATEN	½ TEASPOON SODA
½ CUP MOLASSES	I TEASPOON BAKING POWDER
I CUP SOUR MILK OR BUTTERMILK	I TEASPOON SALT
¼ CUP MELTED SHORTENING	½ CUP SEEDLESS RAISINS
2 CUPS WHEAT FLOUR	

Beat the egg, add molasses, sour milk and shortening.

Sift the wheat flour, measure, add soda, baking powder and salt and sift into the liquid. Pour raisins on top and stir up quickly all at once. Bake in a bread pan at 350° for 45 minutes.

Quick Coffee Cake

I ½ CUPS WHEAT FLOUR	½ CUP BROWN SUGAR
½ TEASPOON SODA	¼ CUP SHORTENING
I TEASPOON CREAM OF TARTAR	I EGG BEATEN
¾ TEASPOON SALT	½ CUP MILK

Topping: 2 TABLESPOONS MELTED BUTTER
4 TABLESPOONS BROWN SUGAR
I TABLESPOON WHEAT FLOUR
CINNAMON

Sift the wheat flour, measure, add the other dry ingredients and sift again. Blend in the shortening, then add the egg and milk beaten together. Mix quickly with a spoon then beat slightly.

Spread in a round or square cake pan (the pan should be 1½ inches deep). Cover the top with the melted butter, then sprinkle on the brown sugar, wheat flour and cinnamon mixed together. Bake 20 minutes at 425°.

Yorkshire Pudding

2 EGGS BEATEN	½ TEASPOON SODA
I CUP BUTTERMILK	¼ TEASPOON SALT
I CUP CORN MEAL	

Beat all together with an egg beater for two minutes. Remove meat from roaster, leaving some fat in bottom of the pan. Pour mixture into roaster and bake about 20 minutes at 425°. If the roaster is large it will be necessary to double the recipe.

This may also be baked in a square shallow cake pan with about 4 tablespoons of fat from the roast put in the bottom of the pan. Tip the pan so the fat will grease the sides too. However, we recommend using the roaster if possible as the drippings from the roast make this dish really toothsome.

Boston Brown Bread

½ CUP MOLASSES
2 TABLESPOONS MELTED SHORTENING
1½ CUPS SOUR MILK OR BUTTERMILK
1 CUP OATEN FLOUR

1 CUP WHEAT FLOUR
½ CUP CORN MEAL
¾ TEASPOON SALT
1½ TEASPOONS SODA
½ CUP RAISINS

Mix together the molasses, shortening and sour milk or butter-milk. Sift the wheat and oaten flours, measure, add corn meal, salt and soda and sift into the batter. Add raisins and beat well.

Pour into a greased mold having the mold no more than ⅔ full. Cover closely. Place in a dish of hot water having the water come half way up on the mold. Steam 3 hours.

If placed in several small molds, steam 2 hours.

Vermont Brown Bread

(*with sour milk*)

1 CUP CORN MEAL
1 CUP SIFTED RYE
1 CUP WHEAT FLOUR
2 TEASPOONS SODA

1 TEASPOON SALT
⅔ CUP MOLASSES
2 CUPS SOUR MILK
2 TABLESPOONS MELTED SHORTEN-ING

Combine molasses and sour milk. Add dry ingredients well mixed together, then shortening. Mix well. Pour into well-greased mold ⅔ full. Cover closely and steam 3½ hours, having boiling water come up half way around mold. Half a cup of raisins may be added to the dry ingredients if your taste so dictates a richer dish.

And the Same

(*with sweet milk*)

We believe that the recipe above, using sour milk creates a con-coction of superior merit, particularly for Saturday night with baked beans. We feel it is worthwhile souring the milk especially for this recipe. But, if you want to make it with sweet milk, here's how: Follow the above instructions except use 1½ cups sweet milk in place of the sour, and reduce soda to ½ teaspoon and add 2 teaspoons baking powder.

Doughnuts

2 EGGS BEATEN	½ TEASPOON SODA
I CUP SUGAR	I ½ TEASPOONS BAKING POWDER
2 TABLESPOONS MELTED SHORTEN- ING	I ¼ TEASPOONS SALT
	I TEASPOON NUTMEG
I CUP SOUR MILK OR BUTTERMILK	½ TEASPOON CINNAMON
4 CUPS WHEAT FLOUR	

Add sugar and shortening to beaten eggs, then the sour milk or buttermilk and beat.

Sift wheat flour, measure, add soda, baking powder, salt, nutmeg and cinnamon and sift into the liquid. Mix thoroughly.

Turn out on a board covered with wheat flour. Knead for a half minute, then pat or roll out to ⅜ inch thickness. Cut out with doughnut cutter and fry in deep hot fat.

Buckwheat Hot Bread

I EGG BEATEN	I CUP BUCKWHEAT FLOUR
3 TABLESPOONS MOLASSES	½ CUP WHEAT FLOUR
¼ CUP SOUR CREAM	½ TEASPOON SODA
¾ CUP BUTTERMILK	I TEASPOON CREAM OF TARTAR
	½ TEASPOON SALT

Beat egg, add molasses, sour cream and buttermilk. Sift buckwheat and wheat flours, measure, add soda, cream of tartar and salt and sift all into the liquid. Beat up quickly and bake in an 8 x 8 x 2 in. pan at 350° for 20 minutes. This will burn if the oven is too hot.

Oatmeal Hot Bread

2 EGGS BEATEN	I CUP OATEN FLOUR
¾ CUP MILK	¼ CUP SOY FLOUR
4 TABLESPOONS BROWN SUGAR	¾ CUP WHEAT FLOUR
3 TABLESPOONS SALAD OIL	¾ TEASPOON SALT
	3 TEASPOONS BAKING POWDER

Beat eggs, add milk, brown sugar and salad oil. Sift the flours separately, measure, add salt and baking powder and sift into the liquid. Mix quickly and bake in a buttered 8 inch square pan at 375° for 45 minutes.

New England Cornbread

2 EGGS	1½ CUPS CORNMEAL
I CUP MILK	3 TEASPOONS BAKING POWDER
4 TABLESPOONS MELTED SHORTENING	¾ TEASPOON SALT

Beat the eggs, add milk and shortening (butter or margarine) and beat again. Combine cornmeal, baking powder and salt, then sift into the liquid mixture. Mix quickly and bake in a pre-heated, buttered 8 inch square pan at 375° for 25 minutes. Cut into squares and serve hot.

Date Nut Bread

2 EGGS BEATEN	2 CUPS WHEAT FLOUR
I CUP BUTTERMILK	I TEASPOON SALT
4 TABLESPOONS MELTED SHORTENING	½ TEASPOON SODA
½ CUP CHOPPED NUTS	I TEASPOON CREAM OF TARTAR
I CUP CHOPPED DATES	

Here is a delicious Nut Bread using only dates as a natural sweetening agent.

Beat the eggs. Add the buttermilk and shortening (butter or margarine) melted in a bread pan. Sift the wheat flour, measure, add the salt, soda and cream of tartar and sift again. Add the chopped dates and nuts to this flour mixture and stir until the dates and nuts are coated with flour. Then add the whole mixture to the liquid ingredients. Stir up quickly, just enough to dampen all the dry ingredients. Bake in the bread pan in which you melted the shortening, making sure that the bottom and sides are well buttered. Bake for 50 minutes at 350°.

Oatmeal and Soy Flour Quick Bread

I EGG BEATEN	I CUP OATEN FLOUR
¾ CUP HOT MILK	½ CUP SOY FLOUR
4 TABLESPOONS BUTTER	½ CUP WHEAT FLOUR
⅓ CUP MOLASSES	¾ TEASPOON SALT
½ CUP CHOPPED DATES OR RAISINS	3 TEASPOONS BAKING POWDER

Pour the hot milk over the oaten flour and butter. Stir until the butter is melted. Add molasses and beaten egg. Sift the soy and wheat flours separately, measure into the sifter and add baking powder and salt. Sift into the oatmeal batter. Add dates or raisins (dates are delicious) on top of the flour and mix all together quickly. Bake at 375° in a buttered 8 inch square pan for 45 minutes.

Muffins, Popovers, Crackers

The secret of good light muffins is to avoid overmixing, to sift dry ingredients together once and quickly, and to dash the mixture into muffin pans with speed and dispatch and without yielding to the temptation to beat. Tender, fine texture muffins will result if recipe is followed faithfully.

Wheat Muffins

1⅔ CUPS WHEAT FLOUR
2¼ TEASPOONS BAKING POWDER
3 TABLESPOONS BROWN SUGAR
I TEASPOON SALT

I EGG BEATEN
I⅓ CUPS SWEET MILK
4 TABLESPOONS MELTED SHORTEN-
ING

Mix together the wheat flour, brown sugar, baking powder and salt. Beat egg until light, add milk and slightly cooled shortening. Toss in the wheat flour mixture and stir quickly just enough to dampen the flour. Let stand a few minutes until mixture thickens then drop by spoonfuls into muffin tins and bake (425°) 20 to 25 minutes. Raisins or nuts may be added to the mixture or sprinkled on top of muffins before placing in oven. A small spoonful of jam or jelly may also be placed on each muffin.

Corn Meal and Wheat Muffins

1 CUP CORN MEAL	¾ TEASPOON SALT
¾ CUP WHEAT FLOUR	2 EGGS, WELL BEATEN
2 TEASPOONS BAKING POWDER	1 CUP MILK
2 TEASPOONS BROWN SUGAR	4 TABLESPOONS MELTED SHORTEN-ING

Sift wheat flour once, measure and add the baking powder, salt and sugar . . . now add corn meal and sift once more and mix well. Combine the milk, eggs and shortening and add to dry ingredients stirring only to dampen mixture. Bake in greased pans in hot oven (425°) about 25 minutes or until done. You can make corn bread-sticks by using bread-stick pans. Try splitting the leftover muffins and toasting them, for a morning meal.

Corn Meal and Rye Muffins

1 CUP CORN MEAL	¾ TEASPOON SALT
1 CUP RYE	2 EGGS, WELL BEATEN
2¼ TEASPOONS BAKING POWDER	1 CUP MILK
2 TEASPOONS HONEY	4 TABLESPOONS MELTED SHORTEN-ING

Sift together the dry ingredients. Combine the milk, eggs, slightly cooled shortening and honey. Add to dry ingredients stirring only enough to dampen. Spoon mixture into muffin pans and bake at 425° 20 or 25 minutes or until done. Makes 12 muffins. (Delicious served with honey.)

Rye and Wheat Muffins

I CUP RYE	2 TABLESPOONS BROWN SUGAR
I TEASPOON SALT	2 EGGS, WELL BEATEN
¾ CUP WHEAT FLOUR	I CUP MILK
I TEASPOON CREAM OF TARTAR	¼ CUP SHORTENING, MELTED
½ TEASPOON SODA	

Beat eggs, add milk and shortening. Then add dry ingredients in order given. Mix all together quickly, but do not beat. Bake in muffin tins at 425° about 20 minutes.

Combination Muffins

½ CUP CORN MEAL	2½ CUPS BOILING WATER
I TABLESPOON SUGAR	½ CUP EACH CORN, RYE AND
I TEASPOON SALT	WHEAT FLOURS
I TABLESPOON BUTTER	2 TEASPOONS BAKING POWDER
	I EGG, WELL BEATEN

Cook ½ cup corn meal, sugar, salt, butter and boiling water in top of double boiler one hour. Remove from fire and when cool add balance of corn meal, rye and wheat mixed with baking powder. Fold in beaten egg and bake in iron gem pans or muffin cups for 20 minutes in hot oven, 425°.

Blueberry Muffins

Add ½ to 1 cup of blueberries to corn meal and wheat muffins. Increase sugar to 4 tablespoons.

Nut Muffins

Add ½ cup of chopped nuts to corn meal and wheat muffins or muffins containing rye.

Raisin Muffins

Add ½ cup of raisins to corn meal and wheat muffins or muffins with rye, or place raisins on top of muffins before baking.

Jam Muffins

Before baking corn meal and wheat muffins or muffins with rye in them, place ½ teaspoon of jam, jelly or marmalade on top of each muffin.

Corn Meal Muffins

2 EGGS, SEPARATED	I CUP CORN MEAL
I CUP BUTTERMILK	½ TEASPOON SODA
3 TABLESPOONS SHORTENING	I TEASPOON BAKING POWDER
½ CUP BUCKWHEAT OR WHEAT FLOUR OR RYE	I TEASPOON SALT

Beat the egg yolks with a spoon. Add buttermilk and shortening and stir again. Sift the buckwheat, wheat or rye, measure, add corn meal, soda, baking powder and salt and sift into the liquid. Stir as little as possible to dampen the flour. Fold in beaten egg whites and pour into a warm iron muffin or cornstick pan. Bake at 425° about 20 minutes.

Kernel Corn Cakes

¾ CUP CORN MEAL	2 EGGS WELL BEATEN
2 TEASPOONS BAKING POWDER	½ CUP MILK
1½ TEASPOONS SUGAR	I CUP KERNEL CORN
½ TEASPOON SALT	

Beat eggs until light. Add milk and corn. Mix together the dry ingredients and combine with liquid mixture. Bake in muffin pans in moderately hot oven 425° for about 20 minutes, and you'll have a dish that gives you that luscious natural corn flavor.

Oatmeal Muffins

I CUP HOT BUTTERMILK	½ CUP WHEAT FLOUR
1½ CUPS OATEN FLOUR	½ TEASPOON SODA
I EGG BEATEN	I TEASPOON BAKING POWDER
3 TABLESPOONS BROWN SUGAR	I TEASPOON SALT
¼ CUP MELTED SHORTENING	

Pour the hot buttermilk over the oatmeal, mix, and let it stand while beating the egg. When the oatmeal is thick and slightly cooled add the egg, sugar and shortening, then beat with a spoon for one minute. Sift the wheat flour, measure, add baking powder, soda and salt and sift into the batter. The important thing now is to mix only enough to dampen the wheat. Let the mixture stand in the bowl while greasing the muffin pans. The mixture will start to rise or puff up. Do not disturb this lightness as you gently and carefully fill the tins.

Bake in a hot oven 425° about 20 minutes. These muffins should come out light and crisp with a real oatmeal flavor.

Soy Flour Muffins

2 EGGS BEATEN	1½ CUPS WHEAT FLOUR
1 CUP MILK	½ CUP SOY FLOUR
2 TABLESPOONS MELTED SHORTENING	1 TEASPOON SALT
3 TABLESPOONS HONEY, MAPLE SYRUP OR BROWN SUGAR	3 TEASPOONS BAKING POWDER

Beat the eggs. Add milk, melted shortening (butter or margarine) and sweetening. Beat all together. Sift wheat and soy flours separately. Measure, add salt and baking powder and sift into liquid ingredients. Stir up quickly, just enough to dampen the flour. Spoon the batter into greased muffin pans. Bake at 425° about 20 minutes.

Buckwheat Muffins

1 EGG BEATEN	1 CUP BUCKWHEAT FLOUR
1 CUP MILK	⅛ CUP WHEAT FLOUR
3 TABLESPOONS HONEY	2½ TEASPOONS BAKING POWDER
	½ TEASPOON SALT

Beat the egg, add milk and honey. Mix thoroughly. Sift buckwheat and wheat flours separately, measure, add baking powder and salt and sift all into the liquid. Stir up quickly and bake in buttered muffin pans at 350° for 20 minutes.

Corn Meal Popovers

I ¼ CUPS CORN MEAL   3 EGGS
2 CUPS MILK   ¾ TEASPOON SALT
I TABLESPOON BUTTER

Scald the corn meal with the milk, add butter and salt and beat thoroughly. When cool add the well beaten eggs, beat two minutes with a spoon and pour into hot muffin or cupcake pans.

Bake at 450° for 10 minutes, then decrease to 350° for 20 minutes more. These do not "pop over" to form a hollow popover because the corn meal is too heavy. However, they are raised by steam instead of a leavening agent, as popovers are made, so we give them the name.

Rye or Wheat Popovers

2 EGGS, WELL BEATEN   I CUP RYE, OR WHEAT FLOUR OR A
I CUP MILK   COMBINATION OF BOTH
I TABLESPOON SHORTENING   ½ TEASPOON SALT

Mix beaten eggs, milk and shortening. Sift the rye or wheat, measure, add salt and sift into the liquid. Beat with an egg beater for 2 minutes. Bake for 10 minutes at 500° then decrease to 350° and bake 20 minutes longer.

Popovers lend themselves well to wholegrain flours because they are made without sugar or leavening agents and the finished product reveals the natural, true flavor of the wholegrain. Popovers are raised by steam instead of baking powder or soda. Therefore, it is imperative that they go into a very hot oven (500°) in order that a crust be formed immediately. As soon as they begin to brown the temperature is gradually lowered to 350° to finish the baking.

Crackers

1 CUP WHEAT FLOUR	½ CUP BUTTER
¼ CUP CORN MEAL	5 TABLESPOONS EVAPORATED MILK
1 TEASPOON BAKING POWDER	OR
¼ TEASPOON SALT	4 TABLESPOONS MILK

Sift wheat flour and corn meal. Measure and sift with other dry ingredients into a bowl. Cut in butter with pastry blender until well blended. Add milk to make a stiff dough. Roll out to 1/8 inch thickness. Cut out with biscuit cutter, or cut into squares, triangles, diamonds, etc. Prick with small skewer. Bake on buttered cooky sheet 5 minutes at 375°, or until brown on the bottom. Turn crackers over with a spatula and bake 3 to 5 minutes longer until again brown on bottom.

VARIATIONS: Garlic Salt, Onion Salt, Onion Flakes, Caraway Seed, Dill Seed, Grated Cheese, Poppy Seed may be sprinkled over the crackers and pressed in gently with a spatula before baking.

Wheatmeal Biscuits (Crackers)

¼ CUP SOY FLOUR	¼ CUP POWDERED MILK
¼ CUP CORN MEAL	½ TEASPOON SALT
¼ CUP WHEAT GERM	1 TEASPOON BAKING POWDER
¾ CUP WHEAT FLOUR	¾ CUP BUTTER
⅛ CUP BROWN SUGAR	¼ CUP MILK

Mix together all dry ingredients (sift wheat and soy flours before measuring). Blend butter into dry ingredients using a pastry blender. Add milk and mix thoroughly. Roll out to 3/16 inch thickness. Shape with biscuit cutter. Prick six times with a small skewer. Bake on buttered cookie sheet for 10 minutes at 350°. Turn crackers over and bake 5 minutes more. Cool on wire rack.

Rye Crackers

1 CUP RYE FLOUR	¼ CUP POWDERED MILK
¼ CUP SOY FLOUR	½ TEASPOON SALT
¼ CUP WHEAT GERM	1 TEASPOON BAKING POWDER
3 TABLESPOONS BROWN SUGAR	½ CUP BUTTER
(OPTIONAL)	¼ CUP MILK

Follow Wheatmeal Biscuits directions. These may be rolled to 1/8 inch thickness.

Biscuits, Griddlecakes, Scones & Dumplings

Baking Powder Biscuits

2 CUPS WHEAT OR RYE FLOUR
1 TEASPOON SALT
3 TEASPOONS BAKING POWDER

4 TABLESPOONS SHORTENING
1 CUP MILK (APPROXIMATELY)

Sift wheat or rye flour, measure, add salt and baking powder and sift again. Blend in shortening, then add enough milk to make a moist dough.

Turn onto a wheat floured board, pat or roll out to ½ inch thickness, cut out biscuits and bake at 475° for about 15 minutes.

Sour Cream Biscuits

2 CUPS WHEAT FLOUR	¾ CUP SOUR CREAM
1 TEASPOON SODA	¼ CUP SWEET MILK
1 TEASPOON SALT	

Sift wheat flour, measure, add soda and salt and sift again. Add sour cream and milk to make a moist biscuit dough. A little more milk may be needed. Beat for a minute with a spoon, then roll out ½ inch thick if you wish and cut out with a biscuit cutter, or shape into biscuits with the hands or a spoon.

Bake at 475° 12 to 15 minutes. If you like fat biscuits, place close together in a pan 8 inches square. If you like crisp biscuits, place 1 inch apart on a cookie sheet.

Rye Biscuits

2½ CUPS RYE	1 TEASPOON CREAM OF TARTAR
1 TEASPOON SALT	3 TABLESPOONS BUTTER
½ TEASPOON SODA	1 CUP MILK

Sift the rye, measure, add salt, cream of tartar and soda and sift again. Blend in butter then add enough milk (about 1 cup) to make a moist dough and beat briskly for a minute.

Empty out on a wheat floured board, pat down, turn over and roll out ½ inch thick. Do not knead. Cut into biscuit shape and bake at 475° 12 to 15 minutes.

If you like crisp biscuits, place one inch apart on a cookie sheet. If you like them soft, place close together in a pan 8 inches square.

Soy Flour and Wheat Biscuits

1 CUP SOY BEAN FLOUR	3 TABLESPOONS BUTTER OR
1 CUP WHEAT FLOUR	MARGARINE
1 TEASPOON SALT	2 TABLESPOONS BROWN SUGAR,
4 TEASPOONS BAKING POWDER	HONEY OR MAPLE SYRUP
	¾ CUP ORANGE JUICE

Sift the soy and wheat flours separately, measure into a flour sifter, add salt and baking powder and sift all together into a bowl.

Blend butter or margarine into dry ingredients with a pastry blender until mixture is crumbly. Mix together the orange juice and sweetening. (Maple syrup is best, but brown sugar or honey are also good.) Pour the liquid into the flour mixture. Stir up quickly and beat for a minute or so. The mixture should be quite stiff but still moist. A little more or less orange juice may be needed.

Drop by tablespoonfuls into buttered pan and flatten down with floured spoon or hands to about ½ inch in thickness. Bake at 475° for 15 minutes.

Cream Scones

2 CUPS WHEAT FLOUR	4 TABLESPOONS BUTTER
2 TEASPOONS BAKING POWDER	2 EGGS, BEATEN
½ TEASPOON SALT	⅓ CUP LIGHT CREAM
2 TEASPOONS BROWN SUGAR	

Sift the wheat flour, measure, add baking powder, salt and sugar and sift into bowl. Blend in butter.

Wheat Griddle Cakes

2 CUPS WHEAT FLOUR	2 EGGS, BEATEN
2 TABLESPOONS BROWN SUGAR	1 ½ CUPS SWEET MILK
3 TEASPOONS BAKING POWDER	5 TABLESPOONS MELTED SHORTEN-
1 TEASPOON SALT	ING

Mix together the wheat flour, baking powder and salt. Beat eggs until light, add milk, brown sugar and shortening. Quickly toss in the dry ingredients and stir only enough to dampen the flour. Brown on hot griddle and serve at once with Vermont maple syrup, honey, or applesauce.

Buttermilk Griddlecakes

I EGG, BEATEN
I CUP BUTTERMILK
2 TABLESPOONS MAPLE SYRUP OR
    HONEY
4 TABLESPOONS MELTED SHORTEN-
    ING

⅓ CUP CORN MEAL, AND ⅔ CUP
    WHEAT FLOUR (OR I CUP OF
    ANY WHOLE GRAIN)
½ TEASPOON SODA
I TEASPOON BAKING POWDER
¾ TEASPOON SALT

Beat the egg, add buttermilk, sweetening and shortening. Sift wheat meal, measure, add corn meal, soda, baking powder and salt and sift into liquid. Mix up quickly stirring only enough to dampen the flour. Bake on greased hot griddle.

Corn Meal and Wheat Griddlecakes

⅓ CUP CORN MEAL
⅔ CUP WHEAT MEAL
I TEASPOON CREAM OF TARTAR
½ TEASPOON SODA
¾ TEASPOON SALT

3 TABLESPOONS BROWN SUGAR
3 TABLESPOONS SHORTENING
I EGG
¾ CUP MILK

Cut the shortening into the combined dry ingredients with a pastry cutter. Add the egg and milk beaten together. Cook on moderately hot greased griddle. If the griddle is too hot the cakes will burn and not cook through properly.

Serve with maple syrup or honey or applesauce.

Quick Buckwheat Cakes

I CUP BUCKWHEAT
½ TEASPOON SODA
¾ TEASPOON SALT

½ CUP SOUR CREAM
I CUP MILK, APPROXIMATELY

Mix buckwheat, soda and salt, then add sour cream and enough milk to make a thin batter.

Then remember the rule for buckwheat cakes. A *hot*, well-greased griddle for frying.

Raised Buckwheat Cakes

I CUP BUCKWHEAT	¾ TEASPOON SALT
½ CUP CORN MEAL	2 CUPS BOILING WATER
½ CUP WHEAT FLOUR	½ PKG. YEAST
½ CUP MILK	¼ CUP WARM WATER
	¼ TEASPOON SODA

Scald the corn meal and salt with the boiling water. Beat well and when cool add the buckwheat, wheat flour, and yeast dissolved in the warm water. Let stand overnight, covered. In the morning add the soda dissolved in milk. Cook on hot, well-greased griddle in small cakes quickly.

Save a cup of batter to serve as a starter for the next day.

The two following recipes for buckwheat cakes, sweetened with maple sugar, are old ones which have been used in a Vermont family for several generations.

Maple Sugar Buckwheat Cakes No. 1

I CUP THICK SOUR CREAM	I TEASPOON SODA
I EGG, BEATEN	½ TEASPOON BAKING POWDER
½ CUP MAPLE SUGAR	I CUP BUCKWHEAT FLOUR
	¼ TEASPOON SALT

Beat the egg, add sour cream and maple sugar and mix until the sugar is dissolved. Measure the buckwheat flour, add other dry ingredients and sift into liquid. Mix up quickly to a very thick mixture. Drop by spoonfuls on a greased griddle, leaving space between the cakes. As they cook they spread out. Serve with plenty of butter and maple syrup and you have something really delicious, if you like buckwheat.

Maple Sugar Buckwheat Cakes No. 2

I CUP OF SOUR MILK	⅓ CUP WHEAT FLOUR
½ CUP MAPLE SUGAR	½ TEASPOON BAKING POWDER
¼ CUP MELTED SHORTENING	I TEASPOON SODA
I CUP BUCKWHEAT FLOUR	PINCH OF SALT

Mix together the sour milk, maple sugar and shortening. Sift the dry ingredients into the liquid mixture and cook on a hot griddle until done. Serve with butter and maple syrup or honey.

Kentucky Raised Buckwheat Cakes

I CUP CORN MEAL	½ PKG. YEAST
2 CUPS BUCKWHEAT FLOUR	4 TABLESPOONS MILK
2 CUPS TEPID WATER	½ TEASPOON SODA
2 TABLESPOONS MELTED SHORTENING	2 TABLESPOONS MOLASSES
I TEASPOON SALT	

Dissolve yeast in I cup of the tepid water. Mix together the corn meal, sifted buckwheat flour and salt. Add the yeast, remainder of water and melted shortening, and beat well. Let rise overnight in warm kitchen. In the morning mix together the milk, soda and molasses. Add to the batter and let rise until ready to use. Then bake on hot griddle and serve with honey, Vermont maple syrup or, if you prefer southern style, you could use molasses or sorghum. It is important that the griddle be well greased.

Buckwheat-Rice Griddlecakes

I CUP WARM BOILED RICE	2 EGG WHITES
I CUP MILK	I TABLESPOON MELTED BUTTER
¾ TEASPOON SALT	¾ CUP BUCKWHEAT FLOUR
2 EGG YOLKS	I TEASPOON BAKING POWDER

Mix together milk, rice and salt. Add beaten egg yolks, butter, buckwheat and baking powder. Fold in beaten egg whites. Cook on hot, well-greased griddle.

Batter Cakes

2 CUPS CORN MEAL MUSH*	¾ TEASPOON SALT
½ CUP WHEAT FLOUR	2 EGGS BEATEN
	MILK

Add the beaten eggs to the corn meal mush, then the sifted wheat flour and salt. Mix in enough sweet milk to make a batter not too

See page 69.

thick and not too thin. The amount required depends on the consistency of the mush.

Cook on a fairly hot greased griddle, browning both sides of the cakes. These take a little longer to cook than regular griddlecakes, so allow a little extra time. However, they are well worth waiting for.

Like pones, they may take the place of bread, or be eaten like griddlecakes for breakfast.

Corn Pone

I CUP CORN MEAL	I CUP BUTTERMILK
¾ TEASPOON SALT	I TABLESPOON FAT
⅜ TEASPOON SODA	

Mix together the corn meal, salt and soda. Add buttermilk and mix well.

Melt the fat in a spider and drop the batter by spoonfuls into the spider. Cover and cook over low heat. When one side is brown, turn the cakes over, cover again and continue cooking. It should take 15 or 20 minutes altogether.

These are delicious served with butter in place of bread with any meal, or with maple syrup, honey or applesauce they make a nice breakfast or luncheon dish.

Corn Dodgers

2 CUPS CORN MEAL	I TEASPOON SALT
4 TABLESPOONS SHORTENING	COLD WATER

Blend shortening into corn meal and salt with fingertips or a pastry blender. Add enough cold water to moisten thoroughly having the mixture slightly more moist than piecrust.

Shape into round cakes ½ inch thick. Place on a well-greased griddle and cook slowly until brown and crusty. Turn and cook the other side the same. Serve with lots of butter and maple syrup or honey.

OR, place on a well-greased cookie sheet and bake in a moderate oven until done.

Pioneer's Hoe Cake

2 CUPS CORN MEAL   ABOUT 2½ CUPS BOILING WATER
1 TEASPOON SALT

Mix together the corn meal and salt and then pour over it the boiling water and mix thoroughly to form thick batter. Now grease a griddle with bacon fat and spread out batter into cakes about half-inch thick or a little less. Grease griddle between fryings, or put daub of butter atop each cake before turning. Cook until golden brown and serve hot with butter and Vermont maple syrup or honey.

The name *Johnny Cake* is a corruption of *Journey Cake* for this simple corn meal dish was probably first used by the pioneers when on journeys to frontiers, wars or hunts. Served with fresh country butter, piping hot off the griddle (one person has to stand and fry while the rest eat, and of course you have to eat near the stove), you can't beat this simple concoction for the full-bodied flavor of the stone ground corn meal. If you want to be literal and cook the daubs of batter on a hoe or shovel over a campfire, you will have Hoe Cake.

Hush Puppies

2 CUPS CORN MEAL   1/3 CUP CHOPPED ONION
1 TEASPOON SALT   1 EGG BEATEN
½ TEASPOON SODA   ¾ CUP MILK
1 TEASPOON CREAM OF TARTAR

Mix together the dry ingredients with chopped onion. Add beaten egg, combined with milk, and beat until smooth. You should have a thick mixture which can be formed into oblong shapes or dropped from a spoon. Brown these on both sides in plenty of fat using the same dish in which fish has been fried. These are always eaten at southern fish fries. Some people fry these in deep fat.

Dumplings

2 CUPS WHEAT FLOUR   ¾ TEASPOON SALT
4 TEASPOONS BAKING POWDER   ⅛ CUP MILK (APPROXIMATELY)

Sift wheat flour, measure, add baking powder and salt and sift again. Add milk enough to make a sticky dough. Drop by spoonfuls on top of stew, and steam 12 minutes without lifting the cover.

Corn Meal Dumplings

Follow recipe for Corn Dodgers, see page 50. Drop by spoonfuls on top of stew or greens and steam about 20 minutes.

Luncheon & Supper Dishes

Wheat Waffles

1 ½ CUPS WHEAT FLOUR	2 EGGS
¾ TEASPOON SALT	1 ¼ CUPS MILK
2 TEASPOONS BAKING POWDER	5 TABLESPOONS MELTED SHORTEN-
2 TABLESPOONS BROWN SUGAR	ING

Sift the wheat flour, measure, add salt, and baking powder. Separate eggs and beat yolks until light. Add milk, sugar and slightly cooled shortening. Sift in the wheat flour mixture and stir quickly just enough to dampen the flour. Beat egg whites until stiff, but not dry. Fold into the batter as quickly as possible, stirring no more than necessary. Cook in hot waffle iron until crisp and brown. Serve with Vermont maple syrup or honey.

Corn Meal Waffles

1 CUP CORN MEAL	1 ½ CUPS MILK
1 CUP WHEAT FLOUR	¼ CUP MELTED SHORTENING
2 TEASPOONS BAKING POWDER	3 EGG YOLKS, WELL BEATEN
½ TEASPOON SALT	3 EGG WHITES, STIFFLY BEATEN
3 TABLESPOONS HONEY	

Sift wheat flour, measure, add corn meal, baking powder and salt and sift again. Add egg yolks combined with milk, honey, and shortening. Fold in egg whites. Bake on hot waffle iron and serve with Vermont maple syrup or honey.

Polenta

1 CUP CORN MEAL	¾ TEASPOON SALT
3 CUPS BOILING WATER	¼ POUND GRATED ROMA OR
1 EGG, WELL BEATEN	PROVOLONI CHEESE (ABOUT 1 CUP)

This grand old Italian dish is used widely in America by all races because it is so different.

Stir corn meal into salted water and cook until thick. Remove from fire. Add cheese, stirring until it is all melted, then add beaten egg. Drop by tablespoonfuls on cookie sheet. When cold, brown the cakes in butter on a hot griddle.

Serve plain with butter or with your favorite tomato sauce.

Roast Meat

Any cut of meat suitable for roasting may be rubbed with salt and pepper, and dredged with wheat flour. Pat on 1 teaspoon Worcestershire sauce with fingertips, sprinkle with a mixture of herbs, lay slices of onion over all and bake or pot-roast in the usual manner.

Baked or Fried Fish

Any fish suitable for frying or baking may be rolled in either stone ground corn meal or stone ground wheat flour.

Meat Wheat Casserole

1 CUP CRACKED WHEAT CEREAL*	2 CUPS TOMATOES
(UNCOOKED)	2 TABLESPOONS CATSUP
1 LB. HAMBURG	1 TEASPOON SALT
½ ONION, CHOPPED	1 TEASPOON POULTRY SEASONING
½ CUP BOILING WATER	1 TEASPOON WORCESTERSHIRE
	DASH OF PEPPER

Saute onions and meat in some fat until brown. Pour the boiling water over the cereal and add to the meat. Add the other ingredients and mix thoroughly. Place in a greased casserole and bake at 325° for about 1 hour.

Chicken-Corn Meal Casserole

1 CUP CORN MEAL	1 ½ TEASPOONS SALT
3 CUPS CHICKEN BROTH	1 TEASPOON BAKING POWDER
3 EGG YOLKS	2 TABLESPOONS BUTTER
3 EGG WHITES	1 CUP COOKED CHICKEN, CHOPPED

Pour hot chicken broth over the corn meal and stir until smooth. Cook until thick, stirring to prevent lumps. Remove from fire, add butter and stir until melted.

Beat egg yolks and add to the mixture, then add salt, baking powder and chicken. Fold in beaten egg whites. Pour into a buttered casserole and bake 1 hour at 325°. Serve hot.

Corn Meal-Beef Casserole

1 ½ CUPS BEEF STOCK	2 TABLESPOONS FAT
½ CUP CORN MEAL	⅔ CUP CHOPPED COOKED BEEF
SALT TO TASTE	1 EGG, BEATEN
½ ONION	TOMATOES, SLICED

Cook corn meal in beef stock until thick. Add more salt if necessary. Brown the chopped onion in the fat and add to corn meal mixture with the beef. Add beaten egg last.

Arrange in buttered casserole with alternate layers of sliced tomato.

Bake at 350° for 20 minutes.

See page 71.

Salmon Loaf

I TALL CAN SALMON	I TABLESPOON CHOPPED PARSLEY
½ CUP SAMP CEREAL	2 EGG YOLKS
I CUP MILK	2 EGG WHITES
4 TABLESPOONS MELTED BUTTER	I TEASPOON WORCESTERSHIRE
¾ TEASPOON SALT	SAUCE

Cook the samp in milk for 10 minutes. Add flaked salmon, butter, salt and parsley and mix together well. Then add beaten egg yolks and Worcestershire sauce. Fold in the beaten egg whites and pour the mixture into a buttered casserole. Cover with whole wheat bread crumbs. Set the dish in a pan of hot water and bake 30 minutes in moderately hot oven. Any leftover portions are delicious sliced and browned in butter in a frying pan.

Tuna Loaf

Follow recipe for Salmon Loaf except use tuna fish instead of salmon.

Creamed Lobster

I ½ CUPS MILK	½ CAN LOBSTER
3 TABLESPOONS WHEAT FLOUR	¼ TEASPOON WORCESTERSHIRE
4 TABLESPOONS COLD WATER	SAUCE
¾ TEASPOON SALT	ONION JUICE TO FLAVOR

Combine the wheat flour, water and salt and add to the hot milk. Stir until thickened and the mixture comes to a boil. Add other ingredients and let it rest on back of the stove for about fifteen minutes, that the lobster flavor may permeate the whole thing.

Serve on split Vermont Crackers, warmed in the oven.

Desserts

Steamed Indian Pudding

I CUP CORN MEAL	½ TEASPOON SODA
⅓ CUP WHEAT FLOUR	I QUART HOT MILK
I TEASPOON CINNAMON	I CUP COLD MILK
½ TEASPOON CLOVES	½ CUP MOLASSES
½ TEASPOON NUTMEG	½ CUP MAPLE SYRUP
½ TEASPOON SALT	I CUP CHOPPED SUET
I CUP RAISINS	2 EGGS

Mix the corn meal with the cold milk and stir into the hot milk. Add the suet, molasses, and maple syrup. Beat the eggs until light and add to mixture. Then sift the wheat flour with the soda, spices and salt, mix with raisins, add to mixture and stir all well. Pour into a buttered mold. Cover tight and steam or cook in boiling water 4 hours. Be sure the water is kept boiling constantly and does not boil away.

Serve with boiled cider hard sauce, cream or maple syrup. The pudding may be heated over in the oven making it as fresh and delicious as when it comes from the steaming mold.

Vermont Indian Pudding

7 TABLESPOONS CORN MEAL	½ TEASPOON SALT
I QUART MILK, SCALDED	½ TEASPOON CINNAMON
⅛ CUP MOLASSES	⅛ TEASPOON NUTMEG
⅛ CUP MAPLE SYRUP	

Stir corn meal into scalded milk. Add rest of ingredients. Mix well and turn into buttered casserole. Place in pan of hot water and bake, without stirring, in moderate oven for 3 hours or more. If you want whey, you must be sure and pour in a little cold milk, after it is all mixed.

Baked Indian Pudding with Suet

I QUART MILK	½ TEASPOON GINGER
½ CUP CORN MEAL	½ TEASPOON NUTMEG
¾ CUP BROWN SUGAR	I TEASPOON CINNAMON
I TEASPOON SALT	½ CUP RAISINS
I TEASPOON SODA	¼ CUP SUET, CUT FINE

Scald 3 cups of the milk. Add all other ingredients. Bake 10 minutes in a hot oven. Lower temperature to 300°, stir with a spoon, pour the cup of cold milk on top and bake 1 hour or more.

Samp Indian Pudding

½ CUP SAMP CEREAL, UNCOOKED*	½ CUP CHOPPED NUTS
I QUART MILK	½ TEASPOON SALT
¼ CUP MOLASSES	¼ TEASPOON GINGER
1/3 CUP BROWN SUGAR	½ TEASPOON CINNAMON
½ CUP RAISINS	I TABLESPOON BUTTER

Scald the samp with milk. Add other ingredients, mix well and pour into buttered casserole or deep baking dish, cover and bake

*See page 71.

3 hours at 250°, or set the dish in a pan of hot water and bake at 300°. Serve with boiled cider hard sauce.

Boiled Cider Hard Sauce

I CUP CONFECTIONER'S SUGAR
2 TABLESPOONS BUTTER
BOILED CIDER

Blend butter and sugar until crumbly. Then add just enough boiled cider to dampen the sugar. Beat until smooth, then place in the refrigerator until hard.

Any proportion of sugar and butter may be used and boiled cider added to make the sauce the consistency of a thick icing. There is no reason why this couldn't be used as an icing too . . . on the applesauce wheat cake, for instance.

Firmenty

½ CUP CRACKED WHEAT
I CUP BOILING WATER
3 CUPS MILK
½ TEASPOON SALT

¼ TEASPOON GRATED NUTMEG
⅓ CUP MAPLE SUGAR
½ CUP RAISINS

Sprinkle cracked wheat into boiling water in top of double boiler, stirring until thickened. When water is absorbed add all other ingredients and steam in double boiler for three hours.

Serve hot with cream or boiled cider hard sauce, or maple syrup.

Steamed Oatmeal Pudding

¼ CUP BUTTER OR CHOPPED SUET,
    OR BOTH
½ CUP MOLASSES
I CUP HOT BUTTERMILK
I EGG, BEATEN

I ½ CUPS SCOTCH OATMEAL
½ TEASPOON SODA
I TEASPOON SALT
I CUP RAISINS AND CHOPPED DATES

Crush soda in the palm of your hand and add to the oatmeal and salt. Cut suet or butter into the mixture with a knife. Add hot buttermilk and molasses, then beaten egg and mix well. Add raisins and dates last.

Put this batter into a buttered mold no more than ⅔ full, cover closely, place mold in another dish of boiling water, having the water come half way up on the mold. Steam 2 or 2½ hours adding hot water to the steamer as required.

Serve hot with Boiled Cider Hard Sauce or Steamed Pudding Sauce. This recipe makes a generous amount since it is poor economy to make a small amount in one steaming. However, the pudding is just as good, if not better, warmed in a strainer placed over boiling water for 10 or 15 minutes.

Mother's Suet Pudding

1 CUP SOUR MILK OR BUTTERMILK	1 TEASPOON SALT
1 CUP SUET, CHOPPED FINE	1 TEASPOON CINNAMON
1 CUP MOLASSES	¼ TEASPOON CLOVES
2½ CUPS WHEAT FLOUR	½ TEASPOON NUTMEG
1 TEASPOON SODA	1 CUP RAISINS

Mix together milk, suet and molasses. Sift wheat flour, measure, add other dry ingredients and sift into the batter. Mix well and add raisins.

Pour into a greased mold no more than ⅔ full. Cover closely and steam 3 hours with the water half way up on the mold. Or steam 2 hours in smaller containers.

Serve with Steamed Pudding Sauce.

Mother's Steamed Pudding Sauce

1 CUP SUGAR	2 TABLESPOONS BUTTER
3 TABLESPOONS CORNSTARCH	2 CUPS BOILING WATER
A LITTLE SALT	½ TEASPOON VANILLA

Mix together sugar, cornstarch and salt. Add the boiling water and cook until the sauce is clear, stirring to keep it smooth. Add butter and vanilla last.

Serve hot over steamed puddings.

Southern Souffle

½ CUP CORN MEAL	1½ CUPS MILK
⅛ TEASPOON SALT	GRATED RIND OF HALF LEMON
⅛ CUP SUGAR	4 EGG YOLKS
¼ CUP BUTTER	4 EGG WHITES

Here is a fine dish, equally good as a side dish or dessert. Scald milk and add 2 teaspoons of butter and the salt. Stir in corn meal and cook until the thick paste is smooth. Take off stove and cool. Now cream the rest of the butter with sugar and lemon rind and beat in egg yolks thoroughly. Add the cooled paste. Beat the egg whites until stiff and fold into the mixture. Put into a buttered pudding dish about a 2 inch thick layer of mixture and spread over it a coating of your favorite marmalade, then another layer of mixture and so on, having top layer that of the mixture. Bake about an hour or until done, in medium oven. You can serve wine sauce on it.

Corn Fritters

½ CUP CORN MEAL	I EGG YOLK
¼ CUP BREAD CRUMBS FINE	I EGG, WELL BEATEN
I TABLESPOON SUGAR	I TABLESPOON BUTTER
DASH OF SALT	I CUP MILK
I TEASPOON GRATED LEMON RIND	

Pour sugar and salt into milk and bring to boil in double boiler. Pour in corn meal slowly, constantly stirring. Let it boil gently for 20 minutes. Take off fire and stir in the butter, lemon rind and egg yolk. Turn mixture out onto a dampened bread board, and flatten out about half-inch thick. Cool and cut into round or square pieces as suits your fancy, and dip into the beaten egg, then into bread crumbs. Cook in hot deep fat, until dark brown. Drain and serve with Vermont maple sugar or marmalade.

Pastry or Pie-Crust

I ½ CUPS WHEAT FLOUR	½ CUP LESS I TABLESPOON SHORT-
¾ TEASPOON SALT	ENING
½ TEASPOON BAKING POWDER	⅛ CUP COLD WATER

Sift the wheat meal, measure, add salt and baking powder and sift into mixing bowl. Blend in shortening with a pastry cutter. The mixture should hold together when pinched between fingertips. If it does not, add more shortening. Dampen with cold water from the tap. Do not use ice water. Mix to a dough. Sift a small amount of wheat flour over a pastry cloth using only enough barely to cover the cloth. Cut off enough dough to line a pie plate and roll out on cloth. Turn the pastry over frequently to keep from sticking. Roll as thin as possible and line the pie plate. Fill with any desired pie filling, roll out the upper crust in the same manner. Dampen the rim of the under crust with more cold water and press upper crust into place with a fork dipped in the wheat flour. Cut slits in upper crust before placing on pie. Trim both crusts and bake at 400° around 30 minutes. Bake longer if the filling requires it.

Shortcake

2 CUPS WHEAT FLOUR	4 TABLESPOONS SHORTENING
2 TABLESPOONS SUGAR	½ CUP MILK
3 TEASPOONS BAKING POWDER	½ CUP WATER
I TEASPOON SALT	

Sift wheat meal, measure, add sugar, baking powder and salt' and sift again. Work shortening into dry mixture with a pastry cutter or finger tips. Add liquid all at once, mix up quickly just enough to dampen the dry mixture thoroughly. Spread in round or square cake pan. Bake in hot oven about 15 minutes. When done, cut into squares, or any desired shape, split open, spread with butter and cover with any desired berry or fruit mixture.

Date Bars

¾ CUP WHEAT FLOUR	¼ CUP SHORTENING
¼ TEASPOON BAKING POWDER	2 EGGS, BEATEN
½ TEASPOON SALT	I CUP DATES
I CUP SUGAR	¾ CUP CHOPPED WALNUTS OR PECANS

Blend shortening into the sugar and add the beaten eggs. Sift the wheat flour, measure, add baking powder and salt and sift again into the batter. Put the dates and nuts on top, then mix all together and beat for one minute with a spoon. Spread in well-greased, wheat-floured 8 x 8 inch pan. Bake 15 or 20 minutes at 350°. Cut in any desired shape while still warm.

Brownies

2 SQUARES CHOCOLATE MELTED WITH	¾ CUP WHEAT FLOUR
½ CUP BUTTER OR OTHER SHORTEN-	½ TEASPOON BAKING POWDER
ING	½ TEASPOON SALT
2 EGGS, BEATEN	½ CUP CHOPPED WALNUTS
I CUP SUGAR	I TEASPOON VANILLA

Beat eggs, add sugar, chocolate mixture and vanilla and beat with a spoon. Sift the wheat flour, measure, add baking powder and salt, sift into batter and mix well. Add nuts and beat one minute with a spoon. Pour into shallow pan and bake 15 to 20 minutes at 350°. Cut in squares while warm.

English Fruit Cake

½ POUND BUTTER	¼ CUP BRANDY OR RUM
½ POUND BROWN SUGAR	I POUND SEEDLESS RAISINS
8 EGGS, SEPARATED	I POUND CURRANTS
I ½ CUPS WHEAT FLOUR	½ CUP CITRON
I TEASPOON CINNAMON	I CUP LEMON PEEL
I TEASPOON ALLSPICE	I CUP ORANGE PEEL
I TEASPOON CLOVES	½ POUND DATES
I TEASPOON NUTMEG	I CUP CHOPPED NUTS
½ TEASPOON MACE	½ CUP SIFTED WHEAT FLOUR
TRACE OF GINGER	

Cream the butter with brown sugar. Beat the egg yolks and mix thoroughly with the sugar and butter.

Sift the wheat flour, measure, add spices and sift into the sugar mixture. Mix well and add the brandy or rum. Add the half cup of sifted wheat flour to the fruit and nuts and add to the batter, mixing thoroughly.

Beat the egg whites until stiff but not dry and fold into the mixture.

Pour into well greased, wheat floured dishes filling them no more than ⅔ full. Tie waxed paper over the top and bake at 300° for 3 hours with a large pan of hot water placed in the bottom of the oven. The top of the cake should burst open when done.

Maple Sugar Cake

I CUP SOFT MAPLE SUGAR
    (8 OUNCES)
¼ CUP SHORTENING
I EGG, WELL BEATEN
⅔ CUP MILK

I ¾ CUP WHEAT FLOUR
½ TEASPOON SALT
2 TEASPOONS BAKING POWDER
I CUP CHOPPED NUTS (OPTIONAL)

Blend shortening and maple sugar until smooth. If the sugar is lumpy place over hot water for a few minutes to soften. Add the beaten egg and beat all together well. Add the milk. Sift the wheat flour, measure, add salt and baking powder and sift into the batter. Add the chopped nuts. Beat all together for one minute. Bake in 2 layers 15 to 20 minutes at 350°.

MAPLE FROSTING:—Boil I cup of maple syrup, or I cup of maple sugar plus ¼ cup boiling water, until it will spin a thread. Pour syrup gradually over 2 beaten egg whites, beating constantly. Continue beating until thick enough to spread on a cake and have it stay there.

Sour Cream Devil's Food Cake

⅓ CUP SHORTENING
I ¼ CUPS SUGAR
I EGG, UNBEATEN
3 SQUARES MELTED CHOCOLATE
I TEASPOON VANILLA

½ CUP THICK SOUR CREAM
I CUP SWEET MILK
I ¾ CUPS WHEAT FLOUR
I TEASPOON SODA
½ TEASPOON SALT

Cream the shortening, add sugar all at once and mix with a pastry blender, then beat with a spoon. Add the egg and beat thoroughly, then unsweetened chocolate and vanilla and beat again. Add sour cream and milk and mix until smooth.

Sift the wheat flour, measure, add soda and salt and sift again into the batter. Mix quickly until smooth, then beat with a spoon for one minute by the clock. Pour into 2 greased layer cake pans and bake at 350° about 30 minutes. Spread with any desired frosting.

Applesauce Wheat Cake

½ CUP SHORTENING
I CUP SUGAR
I CUP APPLESAUCE
I ½ CUPS WHEAT FLOUR

I TEASPOON SODA
I TEASPOON CINNAMON
½ TEASPOON GROUND CLOVES
I CUP RAISINS

Cream shortening, add sugar, mix and beat until smooth. Add applesauce and mix thoroughly. Sift wheat flour, measure, add soda and spices and sift into batter. Pour raisins on top of flour, mix all together and beat for 2 minutes. Bake in greased and wheat floured loaf cake pan for 35 to 40 minutes at 350°.

This recipe may also be baked as cupcakes.

Old English Drop Cookies

I CUP BROWN SUGAR
½ CUP SHORTENING
I EGG
½ CUP COLD COFFEE
½ TEASPOON CINNAMON
½ TEASPOON NUTMEG

½ TEASPOON SODA
¾ TEASPOON BAKING POWDER
¾ TEASPOON SALT
I CUP WHEAT FLOUR
I CUP RAISINS
½ CUP NUTS IF DESIRED

Mix in order given, beating egg into shortening and sugar mixture. Sift the wheat flour after measuring. Beat all together for one minute. Drop by spoonfuls on greased cookie sheet. Bake at 375° for 10 to 15 minutes.

Green Mountain Hermits

2 CUPS CORN MEAL
½ CUP BUTTER OR MARGARINE
I ½ CUPS BROWN SUGAR
2 EGGS

½ TEASPOON SODA
¾ TEASPOON SALT
I TEASPOON CINNAMON
¼ TEASPOON CLOVES
⅓ CUP CHOPPED NUTS

Cream butter or margarine; add brown sugar in small amounts at a time. Then add the beaten eggs and beat well. Combine dry ingredients with the chopped nuts and add to first mixture,

beating well. Drop by teaspoonfuls on greased pan. Bake at 375°
until done. Cookies like these can't hurt the kids, even if they raid
the cookie jar and eat their fill.

Oatmeal Cookies

½ CUP HOT MILK	1 EGG, WELL BEATEN
1 CUP SCOTCH OATMEAL	1 CUP WHEAT FLOUR
¼ CUP MELTED SHORTENING	1 ½ TEASPOONS BAKING POWDER
¾ CUP BROWN SUGAR	½ TEASPOON SALT
	½ TEASPOON CINNAMON

Pour hot milk over the oatmeal and add the hot shortening. Mix
well and add the sugar, then the beaten egg.

Sift the wheat flour, measure, add baking powder, salt and cinnamon. Sift into the batter, mix thoroughly and drop by spoonfuls
on a greased cookie sheet. Bake at 375° for 10 to 15 minutes. One-
half cup of nuts and ½ cup of seedless raisins may be added if desired or used for decoration.

Butter Cookies

1 CUP BROWN SUGAR	½ CUP BUTTER
¾ CUP WHEAT FLOUR	2 TEASPOONS BAKING POWDER
¼ CUP SOY FLOUR	½ TEASPOON SALT
¼ CUP CORNMEAL	1 EGG BEATEN
¼ CUP WHEAT GERM	2 TABLESPOONS MILK
½ CUP POWDERED MILK	½ TEASPOON VANILLA

Cut butter into Brown sugar with pastry cutter. Mix together
all other dry ingredients. (Sift wheat flour before measuring, then
add siftings to mixture). Beat together the egg, milk and vanilla.
Add to dry ingredients and mix thoroughly. Shape dough into a
roll 1 ½ inches in diameter. Wrap in wax paper and chill at least
one hour. Slice ¼ inch thick and bake at 400° for 10 to 12 minutes.
Dough may be stored in refrigerator and sliced and baked as
desired.

Aunt Delia's Molasses Cookies

½ CUP BUTTER OR MARGARINE	3¼ CUPS WHEAT FLOUR
1 CUP SUGAR	1½ TEASPOONS SODA
½ CUP MOLASSES	1 TEASPOON SALT
½ CUP MILK	1 TEASPOON GINGER
	1 TEASPOON CINNAMON

Cream butter or margarine, add sugar and cream together. Add molasses and beat thoroughly. Add milk, beat again. Sift wheat flour, measure into sifter, add other dry ingredients. Sift into batter and beat well. Mixture should be a thick drop batter. Drop by teaspoonfuls on greased cookie sheet. Bake at 375° 8 to 12 minutes. When done remove to wire racks immediately to cool.

Coconut Cookies

1 CUP SUGAR	2 TEASPOONS BAKING POWDER
¼ CUP BUTTER OR MARGARINE	¾ TEASPOON SALT
2 EGGS, SEPARATED	1⅓ CUPS SHREDDED COCONUT (4 OZ.)
1¼ CUPS WHEAT FLOUR	½ TEASPOON ALMOND EXTRACT

Cream together sugar and butter. Add egg yolks and beat until light. Sift wheat flour, measure and sift into butter mixture with baking powder and salt. Mix well, add shredded coconut. Beat egg whites until stiff, add almond extract and combine with butter and flour mixture. Drop by teaspoonfuls on buttered cookie sheet. Bake at 400° for 7 minutes, or until cookies are brown on the edges.

Apple Crunch

2 CUPS SLICED RAW APPLES	½ CUP SUGAR
½ CUP RAISINS OR DATES	½ TEASPOON SALT
⅛ CUP BUTTER OR OTHER SHORTEN-	1 CUP WHEAT FLOUR
ING	NUTMEG OR CINNAMON

Put apples and other fruit in buttered casserole. Mix together butter, flour, salt and sugar. Spread on top of fruit and sprinkle with nutmeg or cinnamon. Bake 30 minutes at 350° and serve with hard sauce or cream.

Corn Meal Gingerbread

I CUP CORN MEAL	I CUP SOUR MILK OR BUTTERMILK
½ CUP WHEAT FLOUR	½ CUP MOLASSES
½ TEASPOON SALT	4 TABLESPOONS SHORTENING
I TEASPOON SODA	I EGG
I TEASPOON GINGER	

Heat the molasses; stir in the corn meal, shortening, salt and ginger; cool. Add the milk. Sift the wheat flour, measure, add soda and sift into batter. Add well beaten egg, then mix and beat thoroughly. Pour into a shallow baking pan and bake 25 to 30 minutes at 350°. Serve with whipped cream.

English Hasty Pudding

½ CUP CORN MEAL	I TEASPOON SALT
4 CUPS MILK	½ CUP MAPLE SYRUP
3 TABLESPOONS BUTTER	NUTMEG OR CINNAMON, GRATED

Bring to a boil on high heat, 1 ½ cups of Milk. Reduce heat to low temperature and immediately add the corn meal. Stir constantly with a wire whisk and cook one minute. Remove from heat and add butter, salt and maple syrup. Mix thoroughly. Add remaining 2 ½ cups of milk, mix again. Pour into a buttered casserole, sprinkle top with grated nutmeg or cinnamon. Cover casserole and bake at 250° for 3 hours. Serve warm.

The original English Hasty Pudding was served on Ash Wednesday but our New England version is good any day.

Breakfast Cereals

Corn Meal Mush

The simplest and earliest use of corn meal was to make that famous dish called *Corn Meal Mush*, or *Hasty Pudding*. You put some water into an iron kettle, got it boiling briskly and then stirred in corn meal with a dash of salt until it was thick. Then you cooked it, stirring all the while. This was used by all good old New England families as a breakfast cereal or a supper dish, with milk and maple syrup poured over it. If there was any left over, after it had cooled it was sliced and fried on a hot griddle and served doused with butter and Vermont maple syrup.

Today there are three ways to cook Corn Meal Mush:

1 CUP CORN MEAL
1 TEASPOON SALT
3 CUPS BOILING WATER

METHOD NO. 1:

Sprinkle the corn meal slowly into the boiling salted water, stirring constantly with a spoon or wire whisk. Simmer for ½ hour, stirring almost constantly to prevent burning. Serve hot, as a breakfast cereal or supper dish, with a pat of butter tucked in, and covered with rich milk. Maple syrup, maple sugar or honey may be added. If a softer mush is desired, use more water when cooking.

METHOD NO. 2:

Put the corn meal, salt and hot water (it doesn't need to boil) in the upper part of a double boiler. Stir together. Set over boiling water in the lower part of the double boiler and cook vigorously for ½ to 1 hour. Turn off heat and let stand overnight. In the morning re-heat and serve as in Method No. 1. For a softer mush, add more water to the recipe.

METHOD NO. 3:

Cook the corn meal in salted boiling water over low heat, stirring constantly, until the mush is thick. Place in double boiler and cook at least 30 minutes. It may be cooked this way in the evening and re-heated in the morning as in Method No. 2.

Fried Corn Meal Mush

Cook corn meal mush using Method No. 1, 2 or 3. If making mush especially for frying, less water may be used to make a firmer product.

Pour the corn meal mush into a chilled loaf pan, smooth over the top, and let set overnight to become firm. Or, chill in the refrigerator. When cold, cut into ½ inch thick slices. Brown both sides in butter in a frying pan over moderate heat. They should be cooked until the slices are a golden color and hold together nicely. Serve with butter in place of bread with the main course, or with bacon or sausage for breakfast, or with maple syrup or honey.

Wholegrain Breakfast Cereals

1 CUP CEREAL

1 TEASPOON SALT

4 CUPS WATER

To save time, these cereals can be cooked the night before you want to eat them. In the upper part of your double boiler put the cereal, salt and boiling water and stir with a spoon. Cover and cook from ½ to 1 hour. Then turn off the heat and let the cereal remain over the hot water until morning when you will reheat and serve. This makes four generous servings or six moderate servings.

If you want to stand by and stir, you can cook all these whole grain cereals in about 20 to 25 minutes directly over the heat without a double-boiler, but you must keep stirring the mixture to prevent sticking and burning down.

You will also find, as we have, that there are all manner of delectable combinations to be made with the different cereals, by mixing them, or by adding raisins, prunes, nuts, and other things that are good to eat.

A word of explanation regarding Samp Cereal and Cracked Wheat Cereal. No matter how long you cook them they will never become a gooey soft cereal, but on the contrary, a cereal that is composed of particles like steamed rice, and therefore something that can be chewed.

The original Samp was the name originally given by the Massachusetts Indians to a native corn they pounded into coarse meal with a pestle and mortar, said mortar being a hollowed-out tree stump and said pestle being a large wooden club hitched to a bent-over sapling to lift it up. Samp mortars could be heard up and down Cape Cod and Long Island so that fog-bound sailors often found their way home on the Samp sound beam!

Samp cereal is coarsely ground corn and wheat, blended.

Other wholegrain breakfast cereals are Cracked Rye and Corn Grits, both the same consistency as Cracked Wheat. Crushed Wheat is a finer ground wheat which will cook into a soft cereal for children and older folk. Cereals consisting of the whole kernels of grain, such as Oat Groats, Kernel Wheat and the like will never cook soft and mushy but will be like Samp Cereal.

INDEX

72

9 780374 532611